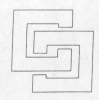

Transport of Dangerous Wastes

European Foundation
for the Improvement of Living and Working Conditions
Loughlinstown House, Shankill, Co. Dublin, Ireland. Tel: (01) 826888 Telex: 30726 EURF EI

This publication is also available in the following languages:

DA - ISBN 92-825-6745-1
DE - ISBN 92-825-6746-X
GR - ISBN 92-825-6747-8
FR - ISBN 92-825-6749-4
IT - ISBN 92-825-6750-8
NL - ISBN 92-825-6751-6

Cataloguing data can be found at the end of this publication.

Luxembourg: Office for Official Publications of
the European Communities

ISBN: 92-825-6748-6
Catalogue number: SY-48-87-234-EN-C

Printed by the European Foundation

PREFACE

In 1984, the European Foundation for the Improvement of Living and Working Conditions decided to launch two research projects on the transport of dangerous goods, substances and waste products in the European Community.

The importance of this subject is beyond doubt. Apart from the economic aspects of the transport of dangerous substances --- both for the industry involved and for the transport sector as such --- the importance of the subject is dictated by preoccupations about safety and environmental protection. These latter points apply particularly to the transport of dangerous waste, and have been brought to public prominence by the unfortunate affair of the "wandering barrels of Seveso".

In order to examine the problems arising from this area, two distinct but complementary research projects were carried out. The first task was to study the legal problems posed by the transport of dangerous goods, substances and waste products. At a later stage, attention was focussed on the specific problems posed by the transport of dangerous waste, especially from the technical point of view.

* * * * *

The present consolidated reports have been drawn up on the basis of national reports describing the situation in the following member states: Belgium, Denmark, Germany, France, Ireland, Italy, Luxembourg, Netherlands and the United Kingdom. These national reports are available separately.

In order to obtain the optimum results from these research projects, close co-ordination was established between the research, the Foundation and the ERCO company in Brussels.

On the basis of the Foundation's guidelines, and in co-operation with the Foundation's project manager, Mr Di Martino, ERCO Sprl drew up a questionnaire addressed to legal experts in the different countries. This questionnaire, and the other questions for the organisation of the work, were subsequently discussed at a co-ordination meeting held in Brussels on 13 and 14 September 1984. This meeting was also attended by the representatives from the relevant international organisations and the EEC Commission.

The final questionnaire which emerged from this process, and which served as the basis for the national reports, appears as an appendix to the legal report.

The same procedure was followed for the research on technical aspects of the transport of dangerous waste; for this part, the Foundation's project manager was Mr Pedersen. After ERCO had drawn up a provisional list of questions, this list was discussed at a co-ordination meeting held in Brussels on 16 and 17 January 1985.

The definitive list of questions which served as the basis for the national reports appears as an appendix to the technical report.

Following this preparatory wok, the national experts drew up their respective reports, in co-operation with the project managers and ERCO.

The consolidated reports which follow derive from the research carried out by their authors, and from the national reports. The definitive version of these dates from November 1985. One of the difficulties encountered in the course of the research was the fact that the regulations in this area are continually changing and developing. Thus, some changes came into force during, and sometimes after, the period of research in certain member states. The new modifications, which consist essentially in a change in the regulations on categories 3, 6.1 and 8 as regards packaging, are mentioned in the preliminary remarks to the national reports. These reports were submitted for comments to the employer and union organisations, as well as to the competent authorities in the respective member states.

* * * * *

As for the consolidated reports, they were evaluated at a meeting held in Brussels on 17 June 1986, attended by representatives of governments, trade unions and employers, as well as representatives of the EEC Commission.

It should first of all be recorded that the research projects were received very favourably by the participants from all quarters, who felt that these studies constitute a useful reference tool for everybody interested in the transport of dangerous goods, substances and waste products.

As regards the observations formulated at the evaluation meeting, the union representatives emphasised the need for harmonisation at Community level between all regulations on questions of safety and health connected with the transport of dangerous materials, substances and waste products. Emphasis was also placed on the Community's transport policy, which ought to be an integral part of the achievement of the internal market by 1992. The policy should contain the following measures:

- all persons in charge of the transport of dangerous materials, substances and waste products, at all levels and during all phases of transport, including the loading and unloading operations, as well as storage between transports, should be informed of the nature and dangers of the materials, substances and waste products involved, and the precautions to be taken in case of accident;

-   failure to ensure a maximum degree of safety during all
    phases of transport, including the loading and unloading
    operations, as well as storage between transports, should be
    made a criminal offence;

-   a solid training programme for all workers, including drivers,
    involved in the transport of dangerous materials, substances and
    waste products, during all phases of transport, should be provided;
    this training, and the qualification of drivers, should be
    harmonised at Community level;

-   transporters engaging in the transport of dangerous materials,
    substances and waste products ought to have very specific
    qualifications and obtain a licence to carry on this business; the
    rules required in this context should be the subject of
    harmonisation at Community level;

-   problems relating to storage between transports and the place of
    unloading ought to be settled at Community level.

The employer representatives largely supported the findings of the
legal report, which stated that dangerous substances and waste should
continue to be treated in the same way as regards safety during
transportation. Admittedly, the question of waste poses specific
problems which demand other measures, but those measures can be
complementary and need not affect existing transport regulations. On
that point, the industrial representatives regretted the fact that the
technical report does not always bring out this difference clearly.

As regards the Community's competence, the participants at the
meeting recognised that the Community undoubtedly has a standing in
this matter, but there were divergent views as to what conclusions
sould be drawn from that fact. Whereas the report calls for the
establishment of a full Community law on the transport of dangerous
goods, substances and waste products, which would harmonise the laws
of member states, the representatives of industry and governments
took a stand against the creation of such a body of legislation. It

would have the effect of adding a system of extra rules to the existing system, and would make the situation even more complicated. For practical reasons, it was also maintained that member states should continue to negotiate directly within the international organisations concerned with establishing transport regulations, but this, obviously, does not rule out prior Community co-ordination among member states.

The representative of the EEC Commission reminded the meeting that the Commission is currently studying the problems of the transport of dangerous goods, substances and waste products, and that its position will be made clear following the work which is currently in progress. Although it is too early to say whether a Community harmonisation will be proposed, it is however clear that the present situation is unsatisfactory. In fact, the law in member states is different as regards transport on a purely national basis, and when international transport is involved, certain bilateral derogation clauses cause complications in the application of agreements. Moreover, some agreements have never been ratified by certain member states. A solution to these discrepancies could be provided by Community legislation.

In the matter of Community provisions on the cross-border transportation of dangerous waste, it was pointed out that Directive 84/631/CEE is still not applied by member states. Some participants, however, agreed with the critical position taken by the authors of the report, who feel that this directive is not a Community measure in spirit, as it reaffirms the existence of frontiers between member states, whereas one of the main aims of the process of European integration is the abolition of those frontiers.

As regards regulations on liability, the industry representatives emphasised that any system of liability in the area of transport must bear in mind that the transporter is fully in charge of the material being transported. Any system of liability that exempted the transporter from such liability would be dangerous, as it would lead to a lesser degree of care by transporters and therefore an increased level of risk. More generally, the introduction of a harmonised system of liability would be a good thing, on condition that it is fair and realistic. In the search for such a system of liability, the current work of international organisations should be fully taken into account.

As well as expressing the sometimes divergent opinions which have just been summarised, the participants called for improvements in the regulations on the transport of dangerous goods, substances and waste products, and there is no doubt that new measures will be introduced progressively in line with the advances made possible by technological development. An example of this kind of improvement is the introduction of compulsory anti-locking systems on heavy vehicles transporting dangerous materials.

In conclusion, it should be pointed out that the opinions and conclusions expressed in the two reports which follow are the sole responsibility of their authors, and should not be taken as reflecting the views of the European Foundation for the Improvement of Living and Working Conditions.

Brussels, September 1986

PART ONE

TRANSPORTATION OF DANGEROUS GOODS, SUBSTANCES
AND WASTE IN THE EUROPEAN ECONOMIC COMMUNITY
- LEGAL SITUATION

PART TWO

TRANSPORT OF NON-NUCLEAR, TOXIC AND DANGEROUS
WASTES - TECHNICAL, SAFETY AND LEGAL ASPECTS
REGARDING PACKAGING AND MEANS OF TRANSPORT

* * * * *

EUROPEAN RESEARCH & CONSULTING sprl

40 SQUARE AMBIORIX, Bte. 7
B-1040 BRUXELLES

TRANSPORTATION OF DANGEROUS GOODS, SUBSTANCES AND

WASTE IN THE EUROPEAN ECONOMIC COMMUNITY

———————————————

LEGAL   SITUATION

W. D. Gehrmann
J. P. Hannequart
P. A. Maier

November 1985

PREFACE

This report has been prepared as part of a project covering the technical and legal aspects of the transportation of non-nuclear dangerous goods, substances and waste.

Its purpose is to give an overall picture of the legal aspects of the complexities of such transportation, to highlight the weak points of the legal system and to suggest points of departure for a possible solution at Community level.

In view of the many problems involved, limits must inevitably be placed on the scope of the subject. Thus transport by air and matters connected with the specific problem of insurance have had to be excluded.

Nevertheless, in Annex II we have included a brief survey of the main problems concerning insurance.

The findings of this research are based on the national reports prepared in accordance with precise questionnaires (see Annex) and on contacts with the international organizations concerned.

This study does not purport to provide an answer to all legal questions which may arise in this field, but our intention has been to give a comprehensive summary of the very complex issues involved in the transportation of dangerous goods, substances and waste.

LIST OF ABBREVIATIONS

ADN	European Recommendations concerning the International Carriage of Dangerous Goods by Inland Waterway
ADNR	European Agreement concerning the Carriage of Goods on the Rhine
ADR	European Agreement concerning the International Carriage of Dangerous Goods by Road
BGB	Civil Code of the Federal Republic of Germany (Bürgerliches Gesetzbuch)
EEC	European Economic Community
CIM	International Convention concerning the Carriage of Goods by Rail
CLN	Convention on Limitation of the Liability of Owners of Inland Waterway Vessels
CMR	Convention concerning Contracts for the International Carriage of Goods by Road
COTIF	Convention concerning International Rail Transport
HNS	Draft Convention on Liability and Compensation in respect of the Carriage of Toxic and Dangerous Substances by Sea
HPflG	Liability Act, governing civil liability for accidents to persons caused by railways and certain dangerous installations in the FRG (Haftpflichtgesetz)
IMDG Code	International Maritime Dangerous Goods Code

LLMC   Convention on the Limitation of Liability in Maritime Claims

NSE   Not specified elsewhere

OBCD   Organization for Economic Co-operation and Development

IMO   International Maritime Organization

UNO   United Nations Organization

PCB/PCT   Polychlorinated biphenyls/Polychlorinated terphenyls

RID   International Regulations concerning the Carriage of Goods by Rail

SOLAS   International Convention for the Safety of Life at Sea

StGB   Penal Code of the FRG (Strafgesetzbuch)

StVG   Road Traffic Act, concerning the conditions of participation in road traffic in the FRG (Strassenverkehrsgesetz)

UNIDROIT   International Institute for the Unification of Private Law

WHG   Water Resources Act, relating to water management in the FRG (Wasserhaushaltsgesetz)

INTRODUCTION

For many years now ever-increasing quantities of goods, substances and waste which constitute a danger to man and his environment have been carried daily by rail, road, inland waterway and sea. These transport operations give rise to growing anxiety.

Various international organizations (for example the Economic Commission for Europe of the United Nations, the OECD and the IMO) have taken the initiative in undertaking studies and establishing conventions designed to make such transport operations safer.

However, the majority of these initiatives are merely a component part of more general programmes relating to toxic and dangerous substances.

In the European Community context, the Environment Action Programmes have emphasized the need to control the disposal of waste and the Council of Ministers has adopted several directives in the field of waste management, a number of which also relate to transport.

More specifically, Directive 84/631/EEC on the supervision and control of transfrontier shipments of hazardous waste falls within this category.

These directives coexist with a complex system of international instruments, some of them long-established (eg the RID and the ADR) and others still under consideration (eg the HNS Convention and the proposed UNIDROIT Convention).

In addition the various national legal systems deal with the matter differently and, to some extent, incompletely.

Thus, different jurisdictions and different levels of legislation encroach upon each other and may create uncertainty and confusion running counter to the requirements of safety and the principles of the Common Market.

For this reason, the European Parliament and the departments of the Commission of the European Communities suggested that the European Foundation should undertake studies in this field to serve as a basis for Community policies and action.

Consequently, Chapter 1 of this report deals with the powers of the EEC and in particular its role in relation to the relevant international conventions, and also with the Community principles which must be observed in the transportation of dangerous goods, substances and waste.

Chapter 2 is devoted to the safety regulations applicable to such transport operations with an attempt to outline the various national approaches and their inadequacies with respect to the existing international conventions.

Chapter 3 concerns liability in the event of accidents occurring during transportation and is to be regarded as supplementing the section on safety regulations.   This discussion also falls within the scope of Directive 84/631/EEC which provides that Community rules on liability must be drawn up for the transportation of dangerous waste.

Lastly, Chapter 4 sets out proposals for harmonization measures which seem essential if transport operations within the Community are to be accomplished smoothly.

1. TRANSPORTATION OF DANGEROUS GOODS, SUBSTANCES AND WASTE:
    THE LEGAL SYSTEMS INVOLVED

Safety in the transportation of dangerous goods, substances and waste
has always been the subject of particular attention. Both at
international and national level, and most recently at Community level,
States have sought to minimize the risk inherent in the dangerous
nature of such operations. They did so at a very early stage by the
adoption of international conventions (the ADR and the RID, for
example), and then by laying down rules applicable in their own States
and, more recently, within the European Economic Community.

At national level, although States long ago laid down certain measures
governing the transportation of dangerous goods and substances, it is
only recently that they have adopted specific measures for "dangerous
waste".

In view of this multiplicity of provisions, given that the transportation of dangerous goods, substances and waste often involves operations
across frontiers, one may well wonder to which legal regime they are
to be subject: to international law, to Community law or to a
succession of different national laws.

The particular features of those provisions will not be considered
in detail, but an attempt will be made to show how they fit together,
that is, to determine which law is applicable and how any contradic-
tions between such provisions may be resolved.

There are three systems of laws, namely:

a. international law;
b. Community law;
c. national law (regional provisions are not dealt with here, nor
    is the relationship between national law and international law,
    since such problems must be settled by reference to the
    constitutional law of each State).

1.1. Relationships between the legal systems

1.1.1. International law and Community law

Community law relating to the transportation of dangerous goods, substances and waste has yet to be created. The existing rules are national and international, namely the ADR, RID and ADNR conventions and the IMDG Code and the various national laws based thereon.

Despite the absence of relevant Community rules at the present time, EEC law must nevertheless be considered from the standpoint of its relationship with international law, since the Community is vested with powers regarding transport (see Articles 74 et seq. of the Treaty).

The EEC Treaty governs the relationship between these two legal systems on the basis of two criteria: first, the attribution of powers as between the Member States and the Community and secondly, the disposition in time of international instruments falling within the scope of Community powers.

International treaties concluded between the Member States and non-member States prior to the Community treaties (eg the RID and the ADR) are not affected by Community law (Article 234 of the EEC Treaty). The position would be otherwise only if those treaties were incompatible with the EEC Treaty, and that is not the case here.

As regards treaties concluded after the EEC Treaty, they must as a rule be concluded by the Community in so far as they fall within the scope of its powers. In order to avoid any incompatibility, Article 228 of the EEC Treaty provides for a priori vetting of such agreements under a procedure whereby the Court of Justice issues an opinion. If the Court's opinion is positive, the Community can enter into a commitment at international level: otherwise, it must either decline to ratify the agreement or else first amend the EEC Treaty under the procedure provided for in Article 236.

Agreements concluded by the Community are binding upon the institutions of the Community and the Member States (Article 228(2) of the EEC Treaty). It should be noted that, despite the Community powers in this matter, the Community has still not ratified the conventions on safety in the transportation of dangerous goods, such as the amended versions of the ADR and the RID.

1.1.2. Community law and national law

The relationships between Community law and national law are based on a hierarchy in which Community law takes absolute precedence.

Thus, any national law on the transportation or management of waste which proved contrary to Community law would be inoperative.

This rule is of primary importance, since if the EEC ultimately adopts transport legislation in this matter, the laws of the Member States must conform with Community law.

1.2. Community powers in the matter of the transportation of dangerous goods, substances and waste

This section deals with the rules on transport in the EEC Treaty and those governing the powers vested in the Community institutions to implement policy in the matter. These powers need to be analysed at both the internal level and the external level.

1.2.1. The Treaty establishing the European Economic Community treats transport as an industry and not merely as a sector associated with several industries; in fact, Title IV of the second part of the EEC Treaty is entirely devoted to transport (Articles 74 to 84). However, those provisions do not apply to all modes of transport: only transport by rail, road and inland waterway fall automatically within their scope, but Article 84 empowers the Council, acting unanimously, to adopt appropriate provisions for sea and air transport.

The Court had occasion to adjudicate as to the scope of the power
vested in the Council in the Schumalla Case (Case 97/78). It took
the view that harmonization of national legislation was an element
of the common transport policy forming part of "the foundations of the
Community". For that reason, Article 75 confers upon the Council a
wide legislative power which enables it to adopt measures of social
policy and road safety.

At the internal level, the introduction of a common transport policy
is therefore a task entrusted to the Community. The transportation
of dangerous goods, substances and waste is a component thereof.

However, in view of the transfrontier nature of transport operations,
the problem soon arose of defining the powers vested in the Community
to conclude external agreements.

Initially, the legal literature recognized only explicit Community
powers, defined by reference to the Treaty, for concluding external
agreements. The provisions in question were principally Articles 111
and 113 (trade agreements) and Article 238 (association agreements).
The Court of Justice overruled that view in a series of judgements
and opinions, two of which relate directly to transport policy, namely
Case 22/70 (the AETR Judgement) and Opinion 1/76.

1.2.1.1. In the AETR Case, the issue was whether it was the Community
or the Member States who had powers to conclude the European Agreement
concerning the Work of Crews of Vehicles engaged in International
Road Transport (AETR), negotiated within the framework of the United
Nations Economic Commission for Europe.

Having regard both to the provisions of Article 75 of the Treaty
concerning a common transport policy and to the implementation of a
Community regulation on that legal basis (Council Regulation No 543/69
of 25 March 1969), the Commission, which maintained that the Community
had powers to sign the international agreement in question, submitted
in particular that an interpretation of Article 75 of the Treaty

enabling the Community to conclude international agreements in the field of transport was in conformity with the concept of the "useful effect" of that provision.   According to the Commission, it would have been unreasonable to have provided for a common policy in a field as extensive as that of transport without giving the Community the appropriate means of action regarding external relations, particularly since, by its very nature, transport frequently has an international aspect and extends beyond the framework of the Community alone.

In its Judgement of 31 March 1971 the Court, after rejecting the theory of the express attribution of power, conceded that Community powers to conclude international agreements might derive implicitly from provisions of the Treaty which did not expressly confer such powers, and from measures adopted, pursuant to such provisions, by the Community institutions.

The Court thereby gave embodiment to the theory of conditional parallel powers.

1.2.1.2. But it was not until Opinion 1/76 that the Court clearly laid down the principle of parallelism of the internal and external powers of the Community.

The Draft Agreement on the Creation of a European Laying-up Fund for Inland Waterway vessels, which the Commission had submitted for an opinion by the Court (under Article 228 of the EEC Treaty) before any pronouncement by the Council on the proposal for a regulation based on Article 75 of the Treaty, clearly raised the question whether, in the absence of any development of secondary law in the sector in question, namely transport policy, Article 75 was, of itself, an adequate legal basis for powers enabling the Community to enter into that international commitment.

The Court gave an unequivocally affirmative answer to that question by confirming, this time in a wholly general fashion, the principle of the parallelism of the internal and external powers of the Community.

It did so in the following terms: "Authority to enter into international commitments may not only arise from an express attribution by the Treaty, but equally may flow implicitly from its provisions." It concluded: "Whenever Community law has created for the institutions of the Community powers within its internal system for the purpose of attaining specific objective, the Community has authority to enter into the international commitments necessary for the attainment of that objective, even in the absence of an express provision in that connection."

The Court stated at a later point in Opinion 1/76 that this conclusion applied not only where the powers had already been exercised within the internal system but also in cases where the internal Community measures were adopted only for the conclusion and implementation of an international agreement, provided that the Community's participation in the agreement was necessary for the attainment of one of the objectives of the Community.

<u>Thus, in all those fields where the Community is vested by the Treaty with powers which are internal in scope, its institutions may elect to exercise their powers either through independent internal action, followed if necessary by external action, or through direct recourse to external action.</u>

1.2.2. However, it is not sufficient to recognize that the Community has powers; it is also necessary to determine to what extent those powers override national powers or whether, on the contrary, the two levels of powers may coexist. It is impossible to reply to this question in general terms. However, the case-law of the Court gives precise guidelines and provides a basis for a solution.

The Court held that the Community had exclusive powers, among other things by reason of the existence of common rules at internal level (AETR Case), by reason of the expiry of a prescribed period (Case 824/79) and in the event of the acknowledged necessity of Community action (Opinion 1/76).

1.2.3. It may be inferred from the foregoing that the Community has exclusive powers at international level regarding the transportation of dangerous goods, substances and waste. Those powers are conferred upon it by the Treaty and necessarily extend to rules governing safety and liability.

1.2.4. The use of such powers at international level

The fact that the Community has exclusive powers provides only an "internal" solution to the problem. It must be possible to give effect to those exclusive powers internationally.

In fact, the exercise of those powers may be complicated by the fact that certain non-member States refuse to recognize the Community as a single party to any international agreement (for example the countries of the Eastern Bloc), and it may also happen that an international organization within which the Community is to put forward its views is unable to accord to the Community the status of member as such.

For instance, the Community has only the role of observer in the UNO (on the Economic and Social Council, which produces recommendations on the transportation of dangerous goods, on the United Nations Economic Commission for Europe, which manages the ADR agreement and the draft ADN agreement, in the International Maritime Organization, which manages the IMDG Code, and on the ADR/RID Committee, which co-ordinates the two agreements governing rail and road transport) and in other international organizations (such as the OECD, although in that case it has a more advantageous status under Additional Protocol No 1 to the Convention concerning the OECD of 14 December 1960).

Another problem may derive from the fact that, despite having exclusive powers, the Community cannot, for internal reasons, fully exercise those powers. This situation arises in particular in cases where the Council fails to act and the Member States do not succeed in arriving at a compromise. Any such failure to act internally affects the capacity of the Community to express its views at international level. In the field of transport, those difficulties are manifest and the Court of Justice recently issued a finding that the Council had failed to act with respect to implementation of a common transport policy (Judgement of 22 May 1985).

But even in such cases, Community powers remain vested solely in the Community, and there can be no question of devolving them to the Member States.

The Court of Justice has nevertheless taken the view that in such situations the Member States may act, but only within very strict limits. In the AETR Case it conceded that the Member States might pursue negotiations but obliged them to act in the interests and on behalf of the Community in accordance with Article 5 of the EEC Treaty.

In Case 804/79 the Court made its position clear by stating that where the Council fails to act, action by the Member States is to be taken in the common interest and in co-operation with the Commission.

In practice, the Community must therefore find a way of giving effect to its powers, whatever the reason for any difficulties which it encounters internationally. It must accordingly take every step to ensure that no international measures accepted by the Member States are inconsistent with existing or future internal measures.

Thus, the Community must speak with one voice with respect to transport agreements and amendments thereto, and the opinion expressed must also take account of future internal Community action. The same applies to the rules governing liability in transport operations which are the subject of draft international conventions (UNIDROIT, IMO).

1.3. The legal regime governing dangerous goods, substances and waste under Community law

Two legal regimes may be applicable, either that relating to the free movement of goods or that relating to the freedom to provide services, the latter being less effective. The two concepts are mutually

exclusive since Article 60 of the EEC Treaty provides that:
"Services shall be considered to be 'services' within the meaning
of this Treaty where they are normally provided for remuneration, in
so far as they are not governed by the provisions relating to freedom
of movement for goods, capital and persons."

1.3.1. Application of the concept of "goods" to dangerous goods, substances and waste

The Treaty gives no definition of "goods". A definition was given
by the Court of Justice in its judgement in Commission v Italy
(Case 7/68): "By goods within the meaning of [Article 9] there must
be understood products which can be valued in money and which are
capable, as such, of forming the subject of commercial transactions".

From that definition it may be inferred that dangerous goods and
substances are indeed "goods" in the legal sense of the term and that
consequently they are covered by the regime created by Article 30
of the EEC Treaty.

The matter of dangerous waste is more complex and a distinction must
be drawn between:

a.   waste which can be recycled to provide secondary raw materials;
b.   waste which it is reasonable to think will be re-usable in the
     near future as a result of technological development; and
c.   waste which may be regarded as permanently non-recyclable.

By virtue of the case-law cited, the first two categories of waste
mentioned fall within the classification of goods; it is only the
third category which raises any real problem.

1.3.2. The specificity of non-recyclable waste

Although non-recyclable waste apparently has no economic value, certain
arguments conduce to its being regarded - from the legal point of
view - as goods:

a.   Commission v Italy (Case 7/68)

In this case, the Court defined goods as "products which can be valued
in money and which are capable, as such, of forming the subject of
commercial transactions"; the second condition is fulfilled in essence.

As regards the first condition, interpretation would appear more
difficult.   On the one hand non-recyclable dangerous waste certainly
has a negative value, but on the other the Court does not state whether
the value of products must be assessed positively or negatively.

Even if it is considered that this fact is not sufficient to bring
non-recyclable waste within the interpretation adopted by the Court,
other principles may militate in favour of such waste being treated
in the same way from the legal point of view.

b.   The principle of legal certainty

If it is taken as a premise that non-recyclable waste is covered only
by the provisions of the Treaty on the provision of services, problems
of legal consistency may arise.

In the first place, the legal regime applicable to certain waste would
change in step with technological developments or indeed according
to the advantages which could be obtained from using it.   A given
type of waste which yesterday was subject to the legal regime governing
services might today be classified as goods and tomorrow - its
attractiveness as a secondary raw material having disappeared as a
result of the substitution of another product or waste - might revert
to its original legal status.   Accordingly there is a risk of some
uncertainty regarding the applicable legal regime.

Matters might also be complicated by a second uncertainty.   Imagine
a case where a Danish company invents a system for recycling or re-using
a type of waste which was previously destroyed or stored; only that
company operates the system and it refuses to grant any licences.   A
French industrialist (in Nice) produces the type of waste in question.
The Danish industrialist offers 100 ECU/tonne as a purchase price

(positive value) but, in view of the dangerous nature of the waste, the transport cost is 200 ECU. An Italian company specializing in the disposal of such waste charges 50 ECU/tonne. Transport costs amount to 50 ECU.

In those circumstances, the producer of the waste is faced with two courses of action which are equivalent from the economic point of view, but his choice of one or the other would involve the application of a different legal regime.

c.   The principle laid down in the judgement of the Court concerning waste oils

In its first judgement in this matter (Case 172/82) and in all subsequent judgements on the same question, the Court held that national legislation whose effect is to ban exports of waste oils for the purpose of disposal in another Member State is contrary to the provisions of the Treaty on the free movement of goods. In its judgement the Court appears to imply that even if the oil has a negative value, it is nevertheless subject to the principle of the free movement of goods.

d.   The principle laid down in the Sacchi judgement and the judgement in Commission v France regarding advertising in respect of alcoholic beverages

In these two cases, the Court established the following principle: national regulations relating to the provision of services may nevertheless fall within the scope of the section of the Treaty governing the free movement of goods where they are of such a nature as to influence movements of goods.

Therefore, if the transportation and disposal of waste are to be viewed in the context of services, may it not be contended that legislation on the transportation and disposal of waste will necessarily have an influence upstream on trade in the products which will give rise to waste? For example, a ban on exports applied in a country where the waste can be neither recycled nor stored will entail a de facto ban on the production of goods which give rise to such waste.

Consequently, it would seem that legislation on the transportation of waste must be analysed in the light of Articles 30 et seq. of the Treaty.

A final argument may be found in

e. The legal basis of the directives concerning waste

All these directives are based on Articles 100 and 235 of the Treaty. This dual legal basis is justified by their purpose, which is set out in the preamble to them.   But the approximation of laws does not necessarily involve recourse to Article 100.   Other Articles prescribe or allow measures to approximate, co-ordinate and harmonize legal provisions in the individual Member States.

The problem of choosing the Article to be used as a legal basis in a given case arises where, by reason of its substantive content, the approximating measure may be regarded as falling equally within the scope of Article 100 and that of another Article.

This is the case in particular where the action taken to approximate laws appears to form part of other Community action.   The question may be asked whether the desired measure should be adopted pursuant to Article 100 or to a provision governing the other action being taken - or, more precisely, whether recourse to the one excludes recourse to the other.

A first solution might be derived from the principle of the _specialia generalibus derogant_ rule.   In this case, Article 100 constitutes the general rule: in fact, it appears within a title of the Treaty headed "Common Rules".   A _lex specialis_ - in this case the rules concerning the provision of services and more particularly Article 63 or 75 - must prevail where it is applicable.

However, if the measures designated to approximate laws go beyond the scope of either one of the Articles, both of them are normally used as the legal basis.

In any event, it would seem that precedence must be accorded to the argument that the Community directives on waste management should have been based on Article 63 if it was considered that they related to the provision of services.

1.3.3. The free movement of goods and the transportation of dangerous goods, substances and waste

On the basis, therefore, that the principle of the free movement of goods is applicable, it must be concluded that dangerous goods, substances and waste produced in one Member State must be allowed to be transported in to any other Member State for the purpose of marketing, treatment, disposal, etc. The only restrictions on such freedom of movement are those connected with the dangerous nature of the goods in question. It is necessarily an exception to a principle to which the Court has always accorded great importance (see Case 172/82 cited above).

Consequently, divergences between the rules of the Member States regarding transportation, whether in the field of safety or in the field of the civil liability of the persons involved or cover of the risks, may lead to distortions in competition.

In the following chapters we shall examine the extent to which the existing international rules on transport already constitute a uniform set of regulations and the extent to which the applicable national provisions are at variance with the common principles.

The answers to these questions will provide a basis for proposals on the action which the Community could take in order to harmonize this complex subject, which falls entirely within its jurisdiction.

2. SAFETY REGULATIONS CONCERNING THE TRANSPORTATION OF DANGEROUS GOODS, SUBSTANCES AND WASTE

2.1. International law

2.1.1. International law and the transportation of dangerous goods, substances and waste

The list of all the international instruments directly and indirectly relating to the transportation of dangerous goods, substances and waste is of impressive length: see Annex 1.

However, a clear distinction must be drawn between the law in force and instruments which have no legal effect, such as conventions which are merely the result of the work of experts or conventions which can not enter into force until certain instruments of ratification have been deposited.

Furthermore, within the context of the law in force a distinction must then be drawn between instruments of binding effect and those which rank only as recommendations.

The latter type of international instrument is fairly frequent in our field of interest; large numbers of technical requirements concerning the transportation of dangerous goods, substances and waste exist only in the form of "codes of good practice".

It may be concluded, at the outset, that the EEC (in view of its own regulatory powers) undoubtedly has a useful role to play by converting certain international provisions into binding law.

Moreover, a major characteristic of the positive international law in this field is its complexity: most of the rules, guiding principles and recommendations are not only highly technical in character but are also constantly being amended and revised. As a result, it is very difficult to determine precisely what requirements apply to a given situation at a given time.

Particular efforts must therefore be made to <u>disseminate the "codes"</u> in force and to ensure that all those professionally engaged in transport operations and also the local authorities, not to mention the public at large, are aware of their meaning and effect.

In analysing the international conventions at present governing the transportation of dangerous goods, substances and waste, a first fundamental distinction to be drawn relates to the mode of transport: road transport, rail transport, transport by inland waterway and transport by sea.

However, <u>multimodal provisions</u> should logically apply in a large number of cases; they would help to clarify the present "legal jungle" and would make it easier for transport operators to carry on their business in a responsible manner.

A good many such provisions already exist in so far as there is a "multimodal source" of standards: the Recommendations of the United Nations Committee of Experts on the transportation of dangerous goods.

Those recommendations, it will be remembered, are brought together in a volume commonly known as the "Orange Book" and relate to the identification and classification of dangerous substances, their packaging, marking, labelling, etc and to the appropriate consignment documents.

2.1.1.1 Rail transport

<u>Transfrontier transportation</u> (including transit) of dangerous goods by <u>rail</u> is subject to the International Regulations concerning the Carriage of Dangerous Goods by Rail (RID).

The RID, which has so far been accepted by 32 States, most of them European, constitutes Annex 1 to the International Convention concerning the Carriage of Goods by Rail (CIM), which was adopted in 1890 and, in its 1984 version, became the Convention concerning International Rail Transport (COTIF).

2.1.1.2. Road transport

For international transportation by road, there exists an addition
to the RID: the European Agreement concerning the International
Carriage of Dangerous Goods by Road (ADR).

The ADR came into existence in 1957 and has so far been ratified by 21
States.   It supplements the Convention concerning Contracts for the
International Carriage of Goods by Road (CMR) adopted in Geneva in 1956.

The ADR and the RID are regularly revised, normally together.   In
principle, therefore, they are harmonized to a satisfactory extent.
However, in view of the relatively high risks involved in road
transport, it may be considered that special measures should be taken
to reinforce the ADR, particularly since in its present form it allows
numerous exceptions or supplementary rules on the part of the
participating States.

2.1.1.3. Transport by sea

It should be noted that transportation by sea is covered by the
International Maritime Dangerous Goods Code (IMDG Code), which was
drawn up by the International Maritime Organization (IMO) within the
framework of the International Convention for the Safety of Life at
Sea (SOLAS Convention).

Although the Code has no legally binding status, it is widely followed
and has been adopted by about 30 countries.

It contains certain specific provisions necessitated by the fact that,
for example, the discharge of gases and   flammable liquids in sea
traffic is much more dangerous for passengers and crew than in the
case of overland or river transport.

The list of "codes" supplementing the IMDG is particularly indicative
of the need for information and training measures, and the need to
devote attention to the matter of due compliance with international
law:

a. emergency instructions for vessels carrying dangerous goods - Emergency cards (1985 edition);

b. guide on urgent medical attention to be given in the event of accidents involving dangerous goods (GSMU, 1985 edition);

c. recommendations on safety in the transportation, handling and storage of dangerous substances in port areas (1983 edition);

d. code of safe practice for solid bulk cargoes (1983 edition);

e. international code for the construction and equipment of ships carrying dangerous chemicals in bulk (IBC);

f. international code for the construction and equipment of ships carrying liquefied gases in bulk (IGC).

2.1.1.4. Transport by inland waterway

In the case of transportation by inland waterway, the Economic Commission for Europe has adopted provisions entitled European Recommendations concerning the International Carriage of Dangerous Goods by Inland Waterway (ADN).

But the recommendations, which are very similar to those of the ADR, are not binding.

However, the European Agreement concerning the Carriage of Dangerous Goods on the Rhine (ADNR) serves very generally as a reference. This Agreement, which was prepared by the States represented on the Central Committee for Navigation on the Rhine (Switzerland, the Netherlands, Belgium, France, the United Kingdom and the Federal Republic of Germany), has been in force since 1972. It differs from the ADN in particular with respect to provisions concerning packaging.

2.1.1.5. The OECD Decision/Recommendation on transfrontier movements of hazardous waste

In order to complete the picture of international law in this field, a legal instrument specifically relating to the transportation of dangerous waste must not be overlooked: the Decision and Recommendation of the OECD Council on Transfrontier Movements of Hazardous Waste.

On 1 February 1984, the OECD Council of the OECD decided that "the member countries shall monitor transfrontier movements of dangerous waste and, for that purpose, shall ensure that the competent authorities in the countries concerned receive appropriate information concerning such waste in due time".

Consequently, the member countries of the OECD are obliged to designate authorities to monitor transfrontier movements of dangerous waste and to require certain persons (to be specified) to furnish information (to be specified) in due time.

The Recommendation concerns principles relating to the public authorities and the various operators engaged by way of trade in the transfrontier shipment of waste.   In the latter connection, it should be noted that under the Recommendation "the producer of dangerous waste is responsible for ensuring that the disposal of the waste (including transport thereof) is carried out in a manner consistent with protection of the environment".

2.1.2. Scope and evaluation of international law concerning the transportation of dangerous goods, substances and waste

In the first place, it should be observed that there are international legal provisions governing the international, but not the internal, transportation of dangerous goods, substances and waste.

Without doubt, as will be seen below, most countries have used the international rules as a guide when laying down their rules for internal transport.   But that does not alter the fact that greater simplicity and legal certainty would be achieved by complete harmonization of the rules for international and national transport operations.

It should also be noted at the outset that there are numerous legal instruments which in principle cover all categories of dangerous goods, substances and waste.

However, attention may be drawn to certain categories which are treated specifically; for example:

a.   nuclear materials;

b.   explosives;

c.   pesticides.

It may be wondered whether, given the present-day socio-economic realities of the transport situation, those categories accurately reflect the need for ad hoc legal provisions (or whether other categories should not be treated in the same way).

In addition, it appears that whilst all the regulations on transportation are structured more or less identically (a list of substances and their classification; general provisions on packaging and transport documentation, followed by quite specific provisions), certain divergences remain regarding the content of the requirements. These divergences are not justified in all cases: a classic example is the fact that sodium chlorite is classified as an oxidizing substance by the ADR and the RID and as a corrosive substance by the regulations for sea transport.   This problem should be solved by incorporating all the Recommendations of the UN Committee of Experts (referred to earlier) in the various international conventions.

More fundamentally, it will be seen that the international rules on the transportation of dangerous goods, substances and waste have as their essential purpose the safety of transport operations.

There is, however, an important exception: the Decision (and Recommendation) of the OECD Council of 1 February 1984, the essential purpose of which is to ensure protection of the environment, regardless of where the dangerous waste is disposed of.

The latter approach is also adopted, of course, in Directive 84/631/EEC on the supervision and control of transfrontier shipments of dangerous waste in the Community.

Consequently, the international transportation of dangerous waste appears to be a specific problem, and a matter deserving of concern as regards, among other things, the safety of transport operations.

"Dangerous waste" must be taken to mean, under the OECD Decision: "any waste other than radioactive waste regarded as dangerous legally defined as dangerous in the country in which it is located or to which it is to be conveyed, by reason of potential risks to man and the environment in the event of accident or of transportation or disposal effected in an inappropriate manner".

Waste itself is defined as "any matter regarded or legally defined as waste in the country in which it is located or to which it is to be conveyed".

It follows, in particular, that an exporting country must concern itself with transfrontier movements of waste which, within its own frontiers, are not regarded as dangerous, if they are so regarded under the law of the importing country.

Fears may of course be entertained that such a solution will not be observed in practice and one can only hope for a true international definition of what constitutes "dangerous waste" in order to facilitate control of its shipment (from an environmental point of view).

From the point of view of safety in transport, it seems that dangerous waste is in fact covered by the existing international conventions, but there are difficulties.

In principle, those conventions treat all substances as dangerous, whether or not they constitute waste, according to their effects; and on that basis they are divided into eight main classes:

1. explosives; 2. gases which are compressed, liquefied, pressurized or refrigerated; 3. flammable liquids; 4. flammable solids and solids liable to spontaneous combustion; 5. oxidizing substances and organic peroxides; 6. toxic and infectious substances; 7. radioactive substances; 8. corrosives. These classes are subdivided into specific items and into generic groups of substances "not specified elsewhere" (NSE), with a code number in each case.

Where a substance is transported with a view to treatment for disposal
or to direct disposal, the United Nations Committee of Experts laid
down, at a meeting in December 1982, that the description of the
consignment should be preceded by the word "waste" (W).   However,
the identification of numerous types of dangerous waste in the form
of mixtures remains a delicate question.

The latter problem was considered at a joint RID-ADR meeting in March
1985.   As a result, new uniform principles were laid down for the
identification of dangerous waste and in particular the procedures
to be followed for the classification of solutions and mixtures.
Waste itself was given the following definition: "all substances,
solutions, mixtures or articles which can no longer be used in their
existing form and which must be transported for treatment, recycling
or disposal by incineration, discharge or any similar process".

It therefore remains to incorporate in positive transport law the
principles concerning dangerous waste recently laid down by the
competent international authorities.

Another matter outstanding is the need to harmonize the international
definition of dangerous waste for the purposes of environmental
protection with that applicable from the point of view of transport
safety rules.

In fact, in so far as several countries have already opted for a
definition of dangerous waste based on degrees of concentration, it
is now important to fix international "quantitative" standards for
the identification of dangerous waste.

2.2. The law of the Member States concerning the transportation of dangerous goods, substances and waste

2.2.1. Sources of law

It goes without saying that the national laws on the transportation of dangerous goods, substances and wastes have many sources. In most cases, very ancient statutes coexist with modern statutes - and harmonization is not always achieved. Moreover, one can only hope, in most countries, for the incorporation of a multitude of sources of law in a single piece of legislation. In the latter connection, it is particularly interesting to note the existence in the Netherlands of a single draft law on the transportation of dangerous substances which is designed not only to bring together the regulations covering different modes of transport but also to reinforce them in all cases.

Moreover, the various laws naturally spawn a multitude of technical requirements.

A reading of certain legislation prompts the question whether the degree of technical complexity adopted is not excessive in a legal context.

In any event, every opportunity to simplify the rules must be exploited (possibly by the introduction of legislation covering the various modes of transport), this being in the interests of legal efficacity.

It should also be noted that (as at international level) the essential purpose of all national rules in this matter is safety in transport operations.

There are, however, a number of enactments which draw direct inspiration from a wish to protect the environment and whose objectives include those such as the reduction of waste at source and the recycling of waste.

In general terms, it is desirable that all transport regulations should include more environmental objectives.

In the latter connection, consideration might be given, for example, to a ban on the transportation of certain types of waste with a view to obliging the producers thereof to undertake its treatment on the spot (and to attempt to prevent its production).

2.2.2. Legal definition of waste

The actual concept of waste is not legally defined in certain Member States of the EEC: Denmark, Ireland and the Netherlands. In the other Member States, the following general definitions are found (in particular):

BELGIUM: products and byproducts which are not used or are unusable (Law of 22 July 1974);

FRANCE/ any residue from a process of production, processing
LUXEMBOURG: or utilization and any substance, material, product or more generally any moveable property which is discarded or intended to be discarded by its holder (Law of 15 July 1975)/(Law of 26 June 1980);

ITALY: any substance or object deriving from human activity or a natural cycle which is discarded or intended to be discarded (Decree of 10 September 1982);

GERMANY: moveable property which the owner wishes to get rid of or whose disposal is mandatory for the protection of the public good (Law of 7 June 1972);

UNITED any object which is "unwanted" or is to be disposed of
KINGDOM: as being broken, worn out, contaminated or otherwise spoiled (Law of 1974).

As a result, the definition of "waste" can hardly be anything other than subjective and/or may vary in space and in time. Consequently, it is difficult to escape a degree of legal uncertainty, and in particular the status of "recyclable waste" raises problems.

With certain exceptions (French law and German law in some cases), the national legal systems seem in principle to exclude "recyclable waste" from the status of waste. However, the legal borderline in that respect is of course very narrow in so far as it depends upon changing economic circumstances: something which is waste at the production stage may subsequently cease to be waste, particularly where a new recycling process becomes available.

It is therefore to be feared that certain "legal loopholes" may be exploited.

To remedy this, it would undoubtedly be useful if national legislatures (like their international counterparts) were to define explicitly the legal status of recyclable waste, by assimilating it to waste properly so called. It is also conceivable for the law on waste to be made more flexible specifically with reference to recycling. (The latter approach was adopted - but perhaps too widely - in Directive 84/631/EEC.)

2.2.3. Legal definition of "dangerous"

"Dangerousness" always depends on particular circumstances.

In national law, the concept of "dangerousness" applies in particular to problems of safety in transport and to the proper management of waste.

From the point of view of safety in transport "that which is dangerous" is normally defined by reference to the Recommendations of the United Nations Committee of Experts on the transportation of dangerous goods.

In the various definitions there nevertheless exist certain different shades of meaning and indeed real divergences (as in the United Kingdom) which of course ought to be eliminated.

As far as rational waste management is concerned, "that which is dangerous" appears to be defined in many ways (sometimes even within the same national legal system).

In BELGIUM: "Toxic waste" is waste "which may present a danger of poisoning for living beings or nature".

The list of such types of waste is defined by the King, in particular "by reference to the toxic substances which they contain, the quantity and concentration of such substances and the activity from which they result".

In DENMARK: "Chemical waste" is waste which falls within five specified categories:

1. fats of animal or vegetable origin
2. halogenated organic compounds
3. non-halogenated organic compounds
4. inorganic compounds
5. miscellaneous waste

In FRANCE: "Waste creating a nuisance" is divided into some 27 types, within five broad categories:

1. waste containing specified substances such as PCBs, solvents, etc
2. radioactive waste
3. paint, oily waste and sludges containing hydrocarbons
4. waste produced by certain technologies, for example coking
5. waste produced by metal-finishing operations

(The Order of 5 July 1983 includes a specific list of "dangerous and toxic waste", whose import is subject to advance declaration).

In LUXEMBOURG: "Toxic and dangerous waste" is waste "containing or contaminated by the substances or materials listed in the Annex to the Regulation of 18 June 1982 in concentrations or quantities liable to present a hazard".

(The list is similar to the list contained in Directive 78/319/EEC but is more precise in that it measures concentrations in mg/kg and gives specific technical requirements for the solubility test).

In ITALY:

"Toxic and dangerous waste" is waste "containing or contaminated by specified substances" (pursuant to Directive 78/319/EEC "including polychlorinated biphenyls and polychlorinated terphenyls and compounds thereof, in quantities and/or concentrations liable to present a danger to health or the environment" (certain detailed changes were made by the Interadministrative Committee on 27 July 1984).

In THE NETHERLANDS:

"Chemical waste" is:

a) waste composed wholly or partly of chemical substances defined by regulation;

b) waste deriving from chemical processes defined by regulation.

In GERMANY:

"Special waste" is waste "which, by reason of its nature, composition or quantity constitutes a particular danger to health or to the quality of water and air or which is particularly explosive or inflammable or contains or may give rise to pathogenic or transmissible infectious agents; such types of waste are specifically listed in Orders". On that basis, account is taken not only of the various substances but also of the various industrial processes which entail the production of waste.

In the UNITED
KINGDOM:

"Special waste" is any waste which:

a) contains one or more of a number of listed substances and which, by reason of the presence of any such substance,

   (i)   constitutes a danger to life (that is to say, ingestion of a dose of 5 cm^3 by a child weighing 20 kg, or inhalation or skin contact or eye contact for 15 minutes, would normally present serious risks),

   (ii) has a flashpoint of 21°C or less;

b) is a medicinal preparation available only on prescription;

c) is a radioactive substance which has dangerous properties other than radioactivity.

Quite apart from the difference in the qualifying adjectives used, it is clear that there are many differences in the national definitions of "dangerous waste". Several criteria are applied:

a.   the type of danger involved (flammability, toxicity, etc);

b.   the generic category of products in question (solvents, medicinal preparations, etc);

c.   technological origin (finishing of metals, etc);

d.   the presence of a specific substance (PCBs, etc);

e.   etc.

The national laws apply these criteria in a variable fashion and, consequently, the listing of dangerous waste is not identical.

In particular, the composition of waste is quantified in some cases and not in others. And even where standards of concentration are fixed, there is no guarantee that they will be similar - quite the contrary.

Consequently, great efforts must still be made to achieve <u>harmonization</u>, particularly to ensure control of the international transportation of waste with a view to protection of the environment "regardless of the place of disposal of the waste". As stated earlier, the EEC directives applicable to dangerous waste (and the OECD instruments) still need to be supplemented by a true international definition of "dangerous waste" including common quantitative standards.

Of course, it would be useful if the legal definition of "that which is dangerous" as far as waste is concerned from the environmental point of view were harmonized with "that which is dangerous" as far as goods and waste are concerned from the point of view of the law on safety in transport.

2.2.4. <u>Identification of the persons subject to obligations</u>

It seems that scant account is taken in the law of most of the EEC Member States of the fact that numerous operators are involved in the transportation of goods and waste.

However, a legal distinction is frequently found between the <u>consignor</u> <u>and the carrier</u>.

In transportation by road, the <u>driver</u> as such also appears in general to be subject to legal provisions, and likewise the <u>master</u> in transportation by sea or inland waterway.

The question therefore arises whether, in wholly general terms, the law should not be <u>made more precise</u> in order to take account of the socio-economic circumstances now prevailing.

It is noteworthy, for example, that the German regulations on the transportation of dangerous substances by road (June 1983 version) identify the following:

a. the consignor;

b. the loader;

c. the carrier;

d. the driver;

e. the driver's mate;

f. the keeper;

g. the entrepreneur or owner of a business responsible for packaging.

In addition, the German regulations on the transportation of dangerous substances by sea differentiate between:

a. the producer or distributor of dangerous substances;

b. the person issuing the bill of lading;

c. the person responsible for loading;

d. the person responsible for management of the goods in the port;

e. the shipowner;

f. the agent;

g. the master or his representative;

h. the person responsible for supervision.

In the context of the management of dangerous waste, there is frequently an intermediary between the producer and the carrier: the collector of dangerous waste.

At first sight, the various legal systems do not, however, accord any particular status to such intermediaries, except in so far as Belgian law contains the concept of the "acquirer" of toxic waste.

(In France, it should also be noted that the trade of collecting dangerous waste is at present subject to negotiation or approval procedures under the auspices of the Agences de Bassin.)

2.2.5. Obligations for consignors

It is not easy to draw up a list of the existing obligations, under national laws, to which the consignors of dangerous goods, substances and waste are subject.

The structure of the legal systems varies, the sources of law are numerous and in some instances obligations are implicit rather than explicit.   A few general observations will therefore suffice at this stage.

In the first place, it must be emphasized that whilst most of the obligations attaching to consignors are inspired by the provisions of international conventions, that fact does not prevent national legislatures from "completing the picture" in many cases.   They lay down certain specific national obligations while applying the principles of international law to "internal" transport operations; on occasion they even lay down a number of special rules for "international" transport operations, by prescribing that certain conventions are to be complied with.

An in-depth analysis of these divergences would provide a useful basis for the introduction of uniform European regulations.

Traditionally, legal provisions are in force for consignors of goods in general, and those provisions are supplemented for the special case of dangerous goods and of dangerous waste.

Every consignor must normally:

1.   prepare a "transport document" and hand it to the carrier;
2.   deal with stowage and identification of the load;
3.   inform the carrier as to the exact content of the load.

Where the transportation of dangerous goods is concerned, involving a consignment on a certain scale (quantitative threshold varies under individual national regulations, which should therefore be harmonized more closely), the consignor is generally obliged to:

1.   prepare a relatively detailed set of transport documents (including a declaration by him to the effect that the applicable regulations have been observed);

2.   observe a number of specific requirements relating to packaging, marking and labelling of the goods;

3.   provide the carriers with all relevant information regarding precautions to be taken, including instructions to be followed in the event of accident.

Where the transportation of dangerous <u>waste</u> is concerned, the consignor - who is normally the producer - must in general comply with a number of special rules, in addition to those mentioned under the heading "dangerous goods".

1.   he must, according to procedures which vary according to the national law in question, notify certain information to the public authorities;

2.   he must ensure that his waste is dispatched to an appropriate disposal installation.

On the latter point, the extent of his obligation nevertheless remains extremely <u>imprecise</u> in most cases.   Only Belgian law is clear, in so far as it provides that the producer "shall be liable for all damage of whatever kind which may be caused by toxic waste, particularly during the transportation, destruction, neutralization or disposal thereof".

PRINCIPAL OBLIGATIONS OF CONSIGNORS	B	D	F	IR	IT	L	NL	FRG	UK
Obligations applicable to all transportation of dangerous goods									
- transport document must be filled in	+	+	+	+	+	+	+	+	+
- stowage of the load must be supervised	+	+	+	-	+	+	+	+	+
- the load must be identified	+	+	+	+	+	+	+	+	+
- the carrier must be informed of the exact content of the load	+	+	+	+	+	+	+	+	+
- the carrier must be informed of special safety precautions to be taken	+	+	+	-	+	+	+	+	+
- the carrier must be given instructions to be followed in the event of accident	+	+	+	+	+	+	+	+	+
Obligations specific to the transportation of dangerous waste									
- the public authorities must be notified before any transport operation	-		-	-	-	+	-	+	+
- the public authorities must be informed on a regular basis	+	+	-	-	+	-	+	+	-

PRINCIPAL OBLIGATIONS OF CONSIGNORS

(continued)

	B	D	F	IR	IT	L	NL	FRG	UK
- a "trip-ticket" or specific transport document must be completed	-	-	+	+	+	+	-	+	+
- a special register must be kept	-	-	+	+	+	+	-	+	+
- certain types of waste must be transferred to a specified place	-	+	-	-	-	-	-	+	-
- actual dispatch of the waste to the place of disposal must be supervised	+	-	-	-	-	-	-	+	-
- the actual conditions under which the waste is disposed of must be monitored	+	-	-	-	-	-	-	-	-

2.2.6. Obligations for carriers

The legal status of carriers obviously calls for the same preliminary remarks as those made in respect of consignors.

Of the catalogue of main obligations generally attaching to the carriers of dangerous goods, the following may be cited:

1) obligation to use an appropriate means of transport, complying with various technical requirements;

2) obligation to possess an authorization or operator's certificate, at least for certain transport operations (see below);

3) obligation to obtain possession of a "transport document" and, at least for certain transport operations, instructions to be followed in the event of accident (see below);

4) obligation to use qualified personnel and to inform tnem of certain precautions to be taken (particularly as regards lorry drivers);

5) obligation to check whether the "packaging" is damaged;

6) etc.

It may be noted that in most instances special obligations are imposed for the transportation of explosives, and for transportation "in bulk" or "by tanker" as opposed to "in packages".

As regards carriers of dangerous waste in addition to rules (of the ADR type) concerning the carriage of dangerous goods they are generally subject to the following:

1) obligation to inform the public authorities - in various ways - of their activities;

2) obligation to use "special trip-tickets" (however, no requirement as to "trip-tickets" is actually enforced in Belgium or the Netherlands);

3) obligation to handle certain types of waste separately (this obligation derives from Directive 78/319/EEC but is expressly included only in Irish and Luxembourg law).

In some countries (Germany and Italy), carriers of dangerous waste may be required to hold a special licence; in others (Belgium and Luxembourg) a special operator's authorization is required for "the acquisition" or "collection" of dangerous waste.

Carriers of dangerous waste in Germany and Belgium are required to effect insurance.

It is clear that the legal status of carriers, particularly carriers of dangerous waste, should be governed by <u>uniform EEC regulations</u>.

PRINCIPAL OBLIGATIONS OF CARRIERS	B	D	F	IR	IT	L	NL	FRG	UK
Obligations applicable to all transportation of dangerous goods									
- a goods haulage operator's certificate must be held	-	-	-	-	+	-	-	-	+
- the carrier must be specially authorized to carry dangerous goods									
a) in general	-	-	-	-	+	-	+	+	-
b) in certain cases	+	+	+	-	+	+	+	+	-
- specific training conditions as regards the driver must be satisfied									
a) for all dangerous transport operations	-	-	+	+	+	-	-	-	+
b) for certain transport operations	+	+	+	-	+	+	+	+	+
- various technical requirements regarding the means of transport must be complied with	+	+	+	+	+	+	+	+	+
- the packaging of goods (in the broad sense) must be supervised	+	+	+	+	+	+	+	+	+

PRINCIPAL OBLIGATIONS OF CARRIERS (continued)

	B	D	F	IR	IT	L	NL	FRG	UK
- a transport document must be held	+	+	+	+	+	+	+	+	+
- safety instructions must be held	+	+	+	+	+	+	+	+	+
- the authorities must be informed in the event of accident	+	-	-	+	-	-	-	+	-
Obligations specific to the transportation of dangerous waste									
- the public authorities must be notified of each transport operation	-	-	-	-	-	+	-	+	+
- the public authorities must be informed on a regular basis	-	-	-	-	+	+	-	+	+
- "trip-tickets" must be completed	-	-	+	+	+	+	-	+	+
- a special register must be kept	-	-	+	-	+	+	-	+	+
- a special documentation identifying the waste must be held	-	+	-	+	-	-	-	-	-
- certain types of waste must be handled separately	+	-	-	-	-	+	-	-	-
- certain packaging rules must be observed	+	-	)	-	-	+	-	-	+

PRINCIPAL OBLIGATIONS OF CARRIERS

(continued)

	B	D	F	IR	IT	L	NL	FRG	UK
- a special licence must be held as a									
a) carrier of waste	-	-	-	-	+	-	-	+	-
b) acquirer or collector of waste	+	-	-	-	-	+	-	-	-
- financial guarantees must be furnished	+	-	-	-	-	-	-	-	-
- special insurance must be effected	+	-	-	-	-	-	-	+	-

2.2.7. Vocational training of carriers

In most countries, there appears to be hardly any certainty of
training for the carriers of dangerous goods, substances or waste.
Apart from the traditional navigation licence and heavy-goods-vehicle
licence, obligations regarding training are rare.   It should be
noted that the new ADR rules provide that the training of drivers is
to be compulsory for international tanker and large-scale transport.
The training is evidenced by a certificate specifying the class or
classes of dangerous goods for whose transportation the driver has
been trained.

In the Federal Republic of Germany, there is a general aptitude
certificate for operators of haulage undertakings; in the United
Kingdom, there is a special licence for the carriage of goods using
heavy vehicles exceeding 3.5 tonnes; but in general there is only
one real vocational-training system, namely that applied to the
drivers of road tankers, and even then only above a specified
capacity (3000 litres).

The latter training does not appear to be suitable for persons who
engage specifically in the collection and transportation of dangerous
waste.   Moreover, it does not seem (unfortunately) to extend to
drivers' mates or to managerial staff.   Moreover, it would be very
desirable for a compulsory vocational-training scheme to apply also
to the carriage of packages.

It should however be noted that various efforts are being made
regarding vocational training for the carriage of dangerous goods in
certain Member States, for example in the United Kingdom through the
Chemical Industries Association and the Road Transport Industry
Training Board.

2.2.8. Rules on traffic movements

In addition to the very many technical requirements applicable to
the various means of transport (lorries, boats, tankers, containers
and so on), which of course raise problems of "technical inspection",

there are a number of specific rules on traffic in the strict sense for the carriage of dangerous goods. Although in maritime transport most traffic rules are formulated at international level, that is not the case in road transport.

Most national regulations on the carriage of dangerous goods by road lay down:

a.   speed limits (for ordinary roads, motorways and built-up areas);

b.   restrictions on road use on certain days (Sundays and public holidays);

c.   restrictions on the use of certain stretches of road (and certain tunnels).

Powers to adopt such regulations are often decentralized, which may result in the need for co-ordination at a higher level. In addition, rules on the parking of vehicles carrying dangerous loads appear to be much more comprehensive in some countries than in others. The present volume of traffic movements would perhaps justify advance arrangements for "special parking areas" in accordance with common guidelines.

The very acceptance of certain particularly dangerous substances for transportation may be regulated by means of a general ban (over and above a certain quantity) save where special authorization is obtained from the authorities. All German regulations on road transport appear to be formulated on this basis. The question arises whether an absolute ban on transportation should not rather be imposed for certain substances. Certain provisions to that effect already exist under the ADR and the RID.

Moreover, authorization for road transport may be limited to carriage as far as the nearest port or station. German law also contains a rule to this effect (in particular in cases where the total distance to be covered is 200 km or more, where the greater part may be covered by rail or inland waterway and where the goods are or can be containerized).

It would be worthwhile seriously to consider extending this principle of German law to the EEC as a whole.

In certain cases there is <u>an obligation to notify</u> the public authorities <u>in advance</u> of a transport operation: thus, in the Netherlands the supervisory body with responsibility for dangerous substances must normally be informed 24 hours before any overland transportation of dangerous goods.

A similar obligation exists generally (particularly by virtue of the ADR) with respect to the transportation of explosives; it exists in certain countries (Germany, United Kingdom, etc) with respect to the transportation of dangerous waste.   There is no doubt that this obligation should be applied more systematically and be imposed uniformly throughout the EEC.

2.2.9. <u>Rules on loading and unloading operations</u>

Virtually all the Member States of the EEC consider that the operations of loading and unloading are an integral part of a transport operation and must be subject to certain special requirements in view of the specific risks involved.

Under the international conventions (the ADR, ADNR, etc) various "technical" measures, which differ according to the categories of dangerous substances involved, must be taken.

Those measures are normally extended from the sphere of international transport to that of national transport operations.

In addition, a number of supplementary rules sometimes come into play: for instance, in the Netherlands, a loading and unloading permit must be issued by the local authorities in many cases and a public <u>inspector</u> must be notified and, where appropriate, must be in attendance.

In all countries, _explosive_ substances appear to constitute a separate category from the point of view of rules on loading and unloading. However, at first sight, the requirements display certain divergences and there are grounds for conducting a detailed comparative analysis of the law in this matter with a view to harmonization and simplification.

There are explicit rules governing _temporary storage_ in certain Member States (Germany, Italy and the Netherlands), which is treated as a special operation, but that is not the case in others with the result that there are certain regrettable legal uncertainties. It should be noted that whilst the German, Italian and Netherlands legislatures impose the requirement of an authorization for temporary storage, the German legislature also confers an express power on the competent authority to require a specific route to be followed, thus precluding any inappropriate temporary storage. (The German legislature is at present even contemplating a general ban on the temporary storage of dangerous waste before disposal.)

2.2.10 Rules relating to accidents

A legal principle which applies everywhere is that the carrier must be informed by the consignor of any dangers associated with the substances to be carried, including information regarding special measures to be taken in the event of accident.

An obligation as to "safety instructions in the event of accidents" is imposed by the applicable international conventions in this field.   In particular, the ADR agreement lays down a general obligation regarding safety instructions for each product.

However, the international rules do not always appear to be implemented in internal law (as for example in the case of sea transport in Italy).

In addition, the international rules concerning safety instructions are not always adopted for "national" transport operations; for instance:

a. in many countries, accidents during transport by rail and inland waterway do not appear to be adequately covered by emergency cards;

b. even in the case of road transport, there are _lacunae_ particularly in rules providing for quantitative exceptions.

In a large number of cases, the prescribed safety instructions appear still to be _too vague_, particularly as regards the special instructions to be followed in the event of an accident.

Special "Hazchem" panels (optimized by the industries involved) exist, but their use is binding only in the United Kingdom and, to some extent, in France and Germany.

Moreover, the rules on accidents are rendered more complicated in many countries by special regulations for particular categories of products (explosives, etc) or waste.

Consideration could therefore usefully be given to imposing a specific obligation upon the producers of waste to provide information.

On this point, however, attention must be drawn to the absence of _specific emergency cards_: the existing emergency cards have traditionally been drawn up for useful substances, not for waste.

Quite apart from the obligations regarding "instructions" in case of accident, it should be noted that:

a. the United Nations system identifying the types of danger (the Kemler code) is generally applied, except in the United Kingdom;

b. various manuals and guides are published (such as the "emergency action manual" published by the Italian Association of Motorway Operating Companies, and the "Chemsafe Manual" published in the United Kingdom by the Chemical Industries Association).

In practice, all countries have set up an "emergency intervention service" and also a "consultation service" for the most difficult cases. Co-ordination between these services (even if limited to greater exchange of information) could without doubt be improved.

2.2.11. Obligations concerning documentation

The obligations regarding documentation imposed by the national legal systems for the transportation of dangerous goods, substances and waste are generally of three types:

a. obligation for a consignment note - bill of lading;
b. obligation for instructions regarding measures to be taken in the event of accident;
c. obligation for inspection certificates.

As regards the last category, there is a rather wide range of documents (varying from country to country): ADR training certificates for certain drivers; an approval document (yellow card) for certain vehicles; seaworthiness certificates or road-use permits, and so on.

All these inspection certificates, which provide an essential means of ensuring constant supervision by the public authorities, certainly need to be harmonized (standardized and simplified).

Specifically with regard to waste, "trip-tickets" designed to ensure control "from the cradle to the grave" exist in most countries of the Community - although not in Denmark and the Netherlands. However, the way in which the system is applied varies very greatly, pending effective implementation of Directive 84/631.

<u>A detailed study should be carried out</u> in order to determine the advisability of making all transport operations involving dangerous goods - or at least the most dangerous operations - subject to the requirement that prior written notification be sent to the authorities.

2.2.12. <u>The role of the public authorities</u>

In all countries, it is clear that the public authorities have the responsibility of monitoring transport operations involving dangerous goods, substances and waste.

This involves the issue of certain prior authorizations and the adoption of emergency measures in the event of accident.

There is no doubt that the national supervisory authorities intervene to differing extents, but the greatest difference lies perhaps in the way in which the relevant powers are shared.   In general these powers are shared between a large number - excessively so - of public authorities.   And in certain countries (Belgium and Italy) there is no doubt that <u>heated disputes regarding</u> the overlapping of powers as between national level and regional level, and even local level, impair the effectiveness of public control.

MAIN TASKS TO BE PERFORMED BY THE PUBLIC AUTHORITIES

PRINCIPAL COMPETENT AUTHORITIES	TRANSPORTATION OF GOODS	TRANSPORTATION OF WASTE
BELGIUM		
- Ministry of Communications (Transport Administration)	- ensuring compliance with general regulations	- ensuring compliance with specific rules
	- approval of vehicles	
	- approval of bodies responsible for inspection of packaging	
(Marine Administration)	- maritime policing	
(Department responsible for Navigable Waterways)	- supervision of inland navigation	
- Ministry of Employment and Labour (Safety Administration)	- ensuring compliance with rules on gas containers	
(Hygiene Administration)		
- Ministry of Economic Affairs (Explosives Department)	- ensuring compliance with rules on the transportation of explosives	
- Regional Authorities		- authorizing purchases and imports of toxic waste
		- providing for removal and treatment of waste
- Provincial Governor		- seizing and destroying abandoned toxic waste

DENMARK

- Ministry of Communications (Vessel Inspection Department)
  - ensuring compliance with maritime regulations

- (Dangerous Goods Department)
  - ensuring compliance with general regulations

- National Agency for Environmental Protection
  - preparing identification cards for chemical waste
  - hearing appeals against decisions of municipal authorities

- Municipal Authorities
  - determining specific rules for the transport of chemical waste
  - supervising and inspecting all chemical waste management operations

FRANCE

- National, regional and departmental officers of the Customs authorities and Ministries of Industry, Supply, the Sea, the Environment and the Interior
  - issuing prescribed authorizations and ensuring that they are carried out
  - penalizing any infringement of laws and regulations

- Bureaux of Maritime Affairs in each port
  - ensuring safety of transport operations and all handling operations in ports

IRELAND

- Ministry of Employment
  (Industrial Inspectorate)
  - enforcing all provisions and carrying out related inspections

- Local Authorities
  - planning, organizing and supervising all disposal (and transport) operations involving toxic and dangerous waste

ITALY

- Ministry of Transport and Ministry of Internal Affairs
  - ensuring compliance with general regulations
  - issuing and ensuring compliance with technical requirements applicable to road vehicles
  - adopting regulations regarding qualifications of drivers

- Interministerial Committee for Waste
  - granting authorizations to undertakings transporting toxic and dangerous waste
  - co-ordinating programmes and plans for waste disposal

- Regions
  - planning disposal

- Provinces
  - granting authorizations for waste management
  - carrying out inspections in the strict sense

- Maritime and Port Authorities
  - supervision and adoption of measures in case of accident

LUXEMBOURG

- Ministry of Transport
  - ensuring compliance with general regulations

- Ministry of Public Works
  - inspecting waterways

- Ministry of Agriculture and Public Health
  - pesticide control

- Ministry of the Environment
  - granting authorizations for disposal of dangerous waste and the requisite approvals for collection of such waste
  - receiving trip-tickets

NETHERLANDS

- Ministry of the Environment
  - regulating the marketing of chemicals
  - regulating management of chemical waste and waste oils (rules on notification and authorizations)
  - imposing specified methods of disposal

- Ministry of Transport
  - granting carriers' licences and special authorizations for certain transport operations

- Local Authorities
  - granting permits for unloading

- Public Goods Inspection Department
  - supervising loading and unloading of explosives

FEDERAL REPUBLIC OF GERMANY

- The Bund

  - all prescribed administrative tasks relating to transport by rail, river and sea

  - designating the competent administrations for road transport

- the Länder

- The Authorities designated under the law of the Länder

  - granting transport authorizations

  - imposing responsibility for certain waste

  - providing information on available installations

  - exercising wide powers of investigation and supervision

2.2.13. Specific rules governing international shipments of waste

Certain transfrontier shipments of dangerous waste are governed by specific rules and are in some instances subject to a genuine system of prior authorization.

Such a system is to be found in particular:

a. in German and Danish law with respect to imports of dangerous waste.

However, in a number of cases the applicable legal requirements involve no more than a prior and/or subsequent declaration. For example:

a. under French law, imports and transit of dangerous waste are subject to the requirement of a prior declaration (and the public authorities are entitled to object within a prescribed period);

b. under Belgian law, a declaration must be made within eight days following exports and within a month following imports;

c. under British, Irish and Netherlands law, importers and exporters must inform the authorities by means of the trip-ticket system. (Although under such systems the authorities do not exercise any power of granting authorizations, it seems nevertheless that they may raise objections to certain shipments of waste if certain safety conditions are not satisfied.)

In some countries there is also a system of operators' authorizations which applies to importers and/or exporters, either as such (Belgium) or as "collectors" or "carriers" of dangerous waste (Germany, Italy and Netherlands).

The national rules on transfrontier shipments of dangerous waste basically provide that certain particulars must be notified to the authorities for information or for the issue of an authorization.

The information to be notified is not identical in all countries; in general, however, it appears that information is required on the following:

a.   the characteristics and quantities of the dangerous waste in question;
b.   identification of the various economic operators concerned;
c.   the modes of transport to be used for the waste;
d.   the final destination of the waste.

This information is notified, depending on the country involved, to one or more authorities responsible for supervision.

Where power to grant authorizations is vested in a particular authority, the grant or withholding of such authorizations is, of course, based upon various items of information and the authenticity thereof.

For example, the following in particular are required for the issue of an authorization to import dangerous waste into the Federal Republic of Germany:

a.   a certificate testifying to the capacity of a disposer to take charge of the waste;
b.   a transport permit;
c.   proof of insurance.

In Belgium, operators who import toxic waste by way of trade must, among other things, produce in advance documentary evidence of financial guarantees and of a commitment to effect a standard insurance contract.

Is the agreement of the authorities in the host country a decisive factor in the context of national procedures for authorizing exports of dangerous waste? It seems that the answer is no. In the

Netherlands, the authorities are nevertheless under a duty to inform the authorities in the host country. (Mention may also be made of an agreement for the exchange of information between the BENELUX countries.)

We should also refer to the recent directive, Directive 84/631/EEC, which provides for notification of the non-member countries concerned in the event of the exportation of dangerous waste out of the Community.

3. LIABILITY FOR DAMAGE ARISING FROM THE TRANSPORTATION OF
    DANGEROUS GOODS, SUBSTANCES AND WASTE

The ways in which the laws of the various Member States of the
Community deal with the problems inherent in the reparation of damage,
particularly with respect to third parties and the environment, are
very different.

Admittedly, in all the Member States the ordinary law starts out
from the idea that liability should derive from the concept of fault,
but the courts have followed different paths in modifying this
principle with a view to securing better protection for third parties
and the environment.

The national courts have, in particular, had to deal with the
following matters:

a. Contractual relations between parties involved in a transport
    operation only render those parties subject to the clauses
    of the contract. However, in reality most risks fall outside
    the area covered by the contractual provisions.

b. Tortious or quasi-tortious liability presupposes the existence
    of fault or negligence, for which the burden of proof as a
    rule falls upon the injured party. In many cases, this
    mechanism does not supply appropriate reparation of the damage,
    particularly in cases of damage to the environment.

c. An obligation to pay compensation irrespective of tort and
    fault relates to the risk inherent in the process of transpor-
    tation, with respect either to the load carried or to the
    actual activity of transport. This may give rise to a major
    problem, namely the limitation of such liability.

First, we shall examine the methods, totally different in some
respects, which are applied in the individual Member States of the
Community to solve these problems.

Secondly, we shall summarize the existing or planned solutions at international level and shall endeavour to indicate all the problems which would be presented by the adoption of uniform rules on liability.

3.1 Contractual liability

Normally the carriage of goods is the subject of a contract. The purpose of the contract is to govern relations between the parties involved in the transport operation. Those relations involve in particular the consignor, the loader, the carrier and the consignee, including various persons providing assistance or acting as intermediaries.

At this stage, liability for non-performance or improper performance of the transport operation normally falls upon the carrier, who is bound by an obligation to achieve a result. His liability arises if the goods are damaged during transportation, without the injured party having as a rule to prove any fault on the carrier's part (CMR, Articles 17 and 18, CIM Articles 26 et seq., Brussels Convention Article 18). In the case of sea transport, the international legislature has nevertheless attenuated the carrier's liability by imposing upon him the requirement only of "reasonable care".

On the other hand, the carrier may be exempted from the obligation to make reparation for the damage caused by his failure to discharge his obligation if he proves that the damage in fact resulted from:

a. circumstances which the carrier could not avoid and whose consequences he could not obviate (CIM Article 27.2 and CMR Article 17.2);

b. fault on the part of the creditor (for example where the damage arises from defective packaging, CIM Article 10);

c. orders issued by the rightful claimant (CMR Articles 17.2 and 18.1 and CIM Article 27.2), "rightful claimant" meaning any party to the contract who, personally or through agents, is entitled to give instructions to the carrier;

d. defects inherent in the goods (CMR Articles 17.2 and 18.1 and
    CIM Article 27.2), defects in the vehicle not being regarded
    as a ground for relief from liability (CMR Article 17.3).

The carrier's liability is limited to sums which depend upon the
weight and volume of the goods carried. But the benefit of that
limitation may be withheld from a carrier if he is guilty of wilful
or inexcusable negligence (maritime law) or wilful or serious
negligence (CMR and CIM).

On the other hand, the consignor is liable for the consequences of
non-performance of his own obligations, such as inadequate stowage
or marking of the goods (CIM Article 12.4 and CMR Article 10) or
defective loading where that is his responsibility.

However, the existing international instruments in this field relate
only to goods. All the relevant provisions are based on the idea
that the property to be transported has an economic value for the
consignor and for the consignee and, therefore, that the traditional
concepts of civil law such as the transfer of ownership and the
transfer of risk are applicable. It is for that reason that those
instruments refer to "loss, damage or delay" (for example, CMR
Chapter IV, Article 17).

The concept underlying the transportation of dangerous waste differs
greatly from that described above, since it involves remuneration
paid to one person by another in order that the former shall take
charge of "property" which has no economic value and shall dispose
of it in an appropriate manner. We are therefore far from the
principles of civil law set out above, since the notions of "loss,
damage or delay" are deprived of their meaning. The principal
obligation of the parties must be safe delivery in order to avoid
any damage to third parties or the environment.

In these circumstances, it is not sufficient to limit the carrier's liability to sums based on the value of the property or a flat-rate basis, as in the CMR for example. It must therefore be concluded that the law of contracts for transfrontier shipments of dangerous waste will be subject to the general rules of international private law and therefore essentially, in accordance with the principle of freedom of choice, to those of the national law chosen by the parties (standard contract).

Detailed examination of the national laws in the matter does not fall within the scope of this study.

In general terms, it may be stated that each of the parties is liable for the harmful consequences of his failure to discharge his obligations; this problem has been discussed earlier, particularlv with respect to loading, unloading and preparation of the transport document.

It is in the area of liability in tort, with respect to damage caused to third parties unconnected with the transport contract, or to the environment, that the greatest difficulties arise regarding liability attaching to the transportation of dangerous goods, substances and waste. Such damage may constitute an extremely serious risk and the potential victims are, of course, not in a position to take precautions in advance by means of insurance, for example.

3.2 The rules of national law on liability in tort in respect of the transportation of dangerous goods, substances and waste

Tortious or quasi-tortious liability arises under the internal law of the Member States where one person causes damage to another, whether by his personal act or through the agency of an object in his custody.

Such liability is based on fault (wilful misconduct or negligence) or arises irrespective of fault (strict liability).

An outline is given below of the national legal systems and the ways in which they have developed.

3.2.1. Liability in countries which adopted the civil code of 1804: France, Belgium and Luxembourg

Liability is based on the concepts of custody and fault (Articles 1382 to 1384 of the Civil Code):

> Article 1382: "Any act whatever of a person which causes damage to another shall oblige him, because of the fault from which it has resulted, to make reparation for it."

> Article 1383: "Every person shall be liable for damage caused by him as a result not only of his acts but also of his negligence or lack of care."

> Article 1384, 1: "A person shall be liable not only for damage arising from his own acts but also for damage caused by the acts of persons for whom he is answerable, or things which he has in his custody."

Thus, proof of fault gives rise to liability on the part of the person committing it, and the person who brings about a situation dangerous to third parties is at fault through failing to take the measures necessary to avoid the damage.

With respect to transport, the laws of France, Belgium and Luxembourg recognize a presumption of liability which, besides being very severely interpreted by the Courts, makes it very difficult for the custodian to be exonerated. The custodian is the person who, at the time when the damage arises, is enjoying the powers of use, management and control of the object in question. Normally, the owner of the object is the custodian, but custody is transferred when the object is entrusted to a commercial operator acting as such (transport contract).

As far as the transportation of dangerous substances is concerned, French law draws a distinction between a custodian's responsibility for the structure and responsibility for behaviour.

This distinction derives from a judicial decision: the French Supreme Court of Appeal took the view that the explosion of gas bottles during a transport operation, caused however by a defect in the bottles, was not attributable to the carrier.

The liability of public undertakings (railways) is dealt with differently in France, since in general it falls within the scope of the rules on liability on the part of the State or public authorities (liability for fault or risk, as the case may be).

In addition to the rules of ordinary law, Belgium and France both have special rules governing the liability of producers of dangerous waste.

Article 7 of the Belgian Law of 22 July 1974 on toxic waste provides that any person carrying on a business (agricultural, scientific, commercial, etc) which involves the production of toxic waste shall be liable for any damage caused by such waste, particularly during transportation, even if transportation is effected by someone else.

The French Law of 15 July 1975 on the disposal of waste and the recovery of materials provides that any person who delivers, or causes to be delivered, any noxious waste to any person other than the operator of an approved disposal installation shall be jointly and severally liable with that person for any damage caused by such waste.

3.2.2. Netherlands law

Extra-contractual liability is governed by Article 1401 of the Burgerlijk Wetboek (Civil Code), which provides that any unlawful act which causes damage to another person places the person whose fault gave rise to the damage under an obligation to make reparation for it.   The various elements of that Article have been developed by court decisions.

In particular, an unlawful act is any act or negligence which violates the right of a third party, is contrary to an obligation attaching to the person committing it, is contrary to a public obligation or to public morality or falls short of the care owed to the third party or the property in question.

Since Netherlands case-law is very severe in its interpretation of those provisions, it seems unlikely that any act causing damage to a third party would ever not be held to be unlawful.

3.2.3. Italian law

Article 2050 of the Italian Civil Code, which covers the exercise of dangerous activities, applies to the transportation of dangerous goods, substances and waste.   Italian law thus takes an original approach, in that account is taken, by virtue of an Article which is general in scope, of the dangerous nature of the activity.

It provides that any person who causes damage to others in the exercise of an activity which is inherently dangerous or dangerous by reason of the means employed is under an obligation to make reparation unless he proves that he took all appropriate steps to avoid the damage.   Liability is presumed.

On the basis of that provision, the producer of a dangerous substance may be held liable even if the damage was caused during transportation carried out by third parties.   The acts of the third party do not relieve the producer of his liability if the latter contributed to causing the damage.

Public undertakings such as railways are covered by specific legal
provisions on liability, which do not include presumption of
liability.

3.2.4. United Kingdom law

The usual basis for liability for damage arising out of the
transportation of dangerous goods lies in the tort of negligence.
This consists of a breach of a duty of care which gives rise to
damage which was not expected or desired.

Under United Kingdom law, not every thoughtless or careless act
gives rise automatically to liability. Liability will arise only
if the defendant was under a duty to take care not to injure the
plaintiff. As a result of court decisions, however, this duty of
care has been extended by the "neighbour principle" and the
"proximity" principle. Accordingly, the courts will usually find
that there was a duty of care and that the damage was foreseeable.

A second possible basis of liability is proof that the defendant
is in breach of a duty which the law has laid upon him. The test
here is whether the conduct was that of a reasonable man. It is
an objective test in so far as the personality of the defendant is
not taken into account. However, the standard is not uniform since
regard is had to the fact that the transportation of dangerous goods
presupposes special knowledge and skills. The standards common in
the profession will, therefore, be taken as the point of reference.

It should be noted that in the United Kingdom victims may face many
problems as a result of the fact that there is no principle whereby
a person committing a fault is in every case obliged to pay
compensation for the damage resulting therefrom. The courts may
in fact take account in each individual case of the gravity of the
damage and the likelihood of its occurrence. Certain serious risks
may even be accepted if the measures required to avoid them would
have been disproportionate in terms of cost. In some cases, the
courts have even acceded to the argument that the social or economic

importance of the activity of the defendant should be balanced against the need to compensate the victim.

3.2.5. Irish law

There are no specific legal provisions in Irish law regarding civil liability for the transportation of dangerous goods, substances and waste. There is no relevant case-law.

The general law of liability is the law of torts, a tort being any damage caused by one person to another which is sufficiently serious to require compensation but not serious enough to be regarded as a crime punishable under the penal law.

The matter of liability is also governed by the Civil Liability Act 1961 and the Civil Liability (Amendment) Act 1964.

3.2.6. Danish law

Any person who causes damage to another person by his negligence is, in general, liable to the persons who have suffered damage.

Negligence or misconduct is essential. No complete definition or description of negligence is given in the law. In the transportation of dangerous goods, substances and waste, compliance with the safety regulations laid down by international conventions and transposed into Danish law is of primary importance in determining liability.

The rule of negligence requires that dangerous products should be handled with particular care.

In Danish law there are, however, certain special enactments which either are important in determining liability without specifically regulating it (for example, safety rules, Environmental Protection Act, decrees on waste oils and chemical waste, Marine Environment Protection Act, and so on) or directly regulate liability according to the mode of transport. The second category comprises:

a.   Chapter 12 of the Merchant Shipping Act (based on the 1969
     Convention but wider in scope), which establishes strict
     liability on the part of the owner;

b.   the rules on road transport, which, as a result of court
     decisions, cover all types of damage caused by motor vehicles
     in the event of accident and create a presumption of liability
     which in fact attaches strict liability to the person having
     control of the vehicle;

c.   the rules of the Railway Compensation Act, which attach strict
     liability to the national railway company.

3.2.7. German law

There are two kinds of laws on liability which concern the transportation of dangerous goods, substances and waste.   First, special
rules such as Articles 7 and 18 of the Road Traffic Act concerning
the conditions of participation in road traffic (StVG), the Liability
Act concerning civil liability for railways and certain dangerous
installations (Haftpflichtgesetz) and Article 22 of the Water Resources
Act concerning underground waters, lakes and rivers (WHG), and
secondly the general rule on liability under the ordinary law
contained in Articles 823 et seq. of the BGB (Civil Code).

In general, the special laws prescribe a degree of liability for
risk that is restricted, in many cases, by virtue of the fact that
the amount of the compensation is limited.

The provisions of the StVG prescribe:

a.   liability without fault (Article 7) attaching to the custodian
     of the vehicle, that is to say the person who is using the
     vehicle on his own account and has the disposal thereof,
     regardless of the legal connection between him and the vehicle;

b.   liability for fault (Article 18) with respect to the driver.

The Haftpflichtgesetz (HPf1G) contains rules similar to those of the StVG. There must also be a causal connection between the operation of the train and the accident. The case-law is based on the view that liability is limited to accidents. Such liability for risks attaches only to the party operating the railway; the HPf1G contains no provisions similar to those of the StVG regarding the driver's liability.

Article 22(2) of the Wasserhaushaltsgesetz (WHG) provides that the owner of an installation used for production, processing, storage or transportation is liable for any damage caused by substances which escape from tanks and enter water. This liability for risk is not subject to any limitation of the amount of compensation, but is offset by a limitation of the persons who are entitled to seek compensation (direct users of the water).

Article 823 of the BGB, in its turn prescribes liability for fault; it provides that any person who wilfully or through negligence and without justification violates the life, body, health, freedom or other absolute right of another person shall be obliged to make good the damage caused. That obligation also attaches to any person who infringes a law whose purpose is to protect other people. If the law providing such protection can be infringed without fault, the obligation to make reparation arises only in case of wrongful conduct.

3.2.8. As indicated in the brief outline of the systems of liability in the Member States, liability for damage occurring during the transportation of dangerous goods and waste varies widely according to the legal regimes involved.

However, the following trends are to be noted. Extra-contractual liability is generally based on the principle of fault, involving the breach of a legal obligation or a customary standard of care.

Under the ordinary law, it is for the victim to prove fault. This may present the victims with serious problems in cases of damage caused during transportation, by reason of the complexity of the technical and economic processes involved.

Court decisions in the various countries have endeavoured to "regularize" the situation by introducing modifications regarding the burden of proof and by taking a very strict line regarding the duty of care incumbent upon those engaged in transport operations. Thus, the res ipsa loquitur rule, for example, applies in the majority of the Member States.

Against the background of these developments and the increase in complex economic activity involving a high degree of risk to man and the environment, liability has in a good many cases also begun to be treated as "strict liability".

In the light of this general trend, Belgium (the only Member State so far to do so) has adopted an explicit legal provision providing for strict liability with respect to dangerous waste.

There are equivalent provisions in the environmental law of several other Member States, such as the German Water Resources Act (WHG) and the laws of certain Member States governing relations between neighbours, which provide that damage arises as soon as the balance of interests is disrupted, without any need for fault to be proved (in France and Belgium, the theory of "neighbour problems"; in

Common Law countries, the theory of "nuisances"; In the Federal
Republic of Germany, BGB Paragraph 906, etc).

There is no lack of specific examples of rules on strict liability
governing the exercise of certain dangerous, or abnormally dangerous,
activities.   Thus, in the United Kingdom, the handling of "anything
which is likely to do mischief if it escapes" gives rise to strict
liability under the rule in Rylands v. Fletcher.   That rule, like
the theories on neighbour problems referred to earlier, does not
apply to transport operations generally, but most countries have
established strict liability with respect to traffic accidents
caused by motor vehicles.

The fact that the victim may encounter difficulties regarding proof
and causality is an essential argument favouring the introduction
of strict liability regarding the transportation of dangerous goods,
substances and waste.

The person bearing civil liability is normally the person who <u>caused</u>
damage through his fault.   For that reason, nearly all the theories
of causality are based on the principle that only circumstances
which are susceptible, or may be regarded as reasonably susceptible,
of damage may be regarded as having caused damage.

The consequence of imposing liability for outright fault is that the
victim must prove the causal relationship.   In cases of damage
caused by the transportation of dangerous goods, substances and waste,
such proof will necessarily pose difficulties for the victim since
he will not normally have at his disposal relevant information
enabling him to provide such proof.

This causal connection is particularly important in assessing long-
term or future damage.

Where there is a "presumption of liability" or where the res ipsa
loquitur rule applies, the victim need only provide proof as to the
first link in the causal chain, and it will thereupon be for the
defendant to show that the causal chain was broken if he is to
be relieved of his liability.

The doctrine of strict liability totally reverses the burden of
proof. It is for the defendant to show that no action taken by
him gave rise to the damage, and in any event he can be exonerated
only if he has discharged all his obligations of diligence and
care.

This basic tenet is followed in all the existing and planned
international conventions concerning the matter. We shall refer
again to those conventions in the section dealing with
identification of the persons liable.

3.3. Damage

Any discussion of extra-contractual reparation for damage arising
from the transportation of dangerous goods, substances and waste
must deal in particular with three questions:

a. Types of damage and possibility of reparation
    Damage deriving from the transportation of dangerous goods,
    substances and waste may be divided into two classes: direct
    damage to third parties resulting from the dangerous nature
    of the actual process of transportation and from the substances
    transported, for example fire or explosion, and damage caused
    to the environment which may have more long-term effects.

b. Persons who may claim entitlement to reparation
    Problems arise where the damage affects personal rights which
    are not protected or where it affects the environment.

c. Limitation of liability or compensation.

A brief outline follows of the approaches adopted in the various
Member States, which is based on the traditional distinction between
bodily injury, material damage and non-material damage.

3.3.1. The law on the transportation of goods is concerned
essentially with loss, damage and delay, that is to say matters which
are of interest only to the parties to the transport contract.

In the absence of specific rules, damage to third parties is a matter
covered by ordinary law. This means that the victim will have to
rely on the applicable national system of liability to justify his
right to compensation against the originator of the damage, on the
ground that the latter has committed a fault or because he has
custody of the object in question.

3.3.2. The national courts are unanimous in recognizing entitlement
to compensation where a third party is injured or his property is
damaged by an accident arising from the transportation of dangerous
goods, substances or waste, provided that the damage is direct and

<u>identifiable</u>.   The claim for compensation usually includes a <u>pretium</u> <u>doloris</u> and restitution for all costs arising from injuries, such as expenses for medical treatment and other disbursements.

It should nevertheless be noted that in the Federal Republic of Germany the <u>pretium doloris</u> is granted only where there is liability in tort for fault under Paragraph 847 of the BGB, whereas in Italy a "criminal unlawful act" must have been committed.

As regards compensation for damage extending beyond that described above, the national courts adopt differing approaches.   In France and the BENELUX countries, compensation is normally paid for all the direct adverse consequences of an accident arising from the transportation of dangerous goods, substances or waste, regardless of the nature thereof, such as commercial losses, <u>lucrum cessans</u> or mental anguish in the widest sense of the term.

In the United Kingdom, compensation for damage is usually very comprehensive, but it should be added that in certain limited cases compensation is granted to a victim with the intention of punishing the defendant.   But the possibility of awarding exceptional damages is not the only prerogative available to the United Kingdom courts, since they may, conversely, take the view that the circumstances of the defendant are such that full compensation cannot be granted to the victim.

Fairly extensive compensation is also usually granted in cases of loss or reduction of earning capacity.   Economic loss, on the other hand, used to be recognized only in cases where it was accompanied by physical injury;   however, it was held in a recent decision that compensation ought to be paid for an economic loss pure and simple (Junior Books Ltd v. Veitchi Ltd, 1982).

The courts in the Federal Republic of Germany, for their part, adopt a much more restrictive approach. Paragraph 253 of the BGB, for example, provides for compensation for non-material damage only in exceptional cases. However, more recently there has been a clearly identifiable trend towards the "commercialization" of non-material damage.

This tendency has also emerged in Italy where, in addition to the compensation traditionally awarded for damage, compensation for "biological damage" has been introduced, its purpose being to protect physical and psychological integrity in cases of "unfair prejudice".

3.3.3. However, the national courts are much more reserved with respect to compensation for pollution damage (damage to the environment). Such damage, which affects the community as a whole, is not in principle recognized as qualifying for compensation, in view of the personal nature of the right to compensation.

Provided that such damage can be seen to be direct and certain and provided that someone can enforce his rights in respect of the damaged area (generally governments or municipal authorities), compensation is payable in accordance with the traditional criteria (damage to persons or property). It seems that a number of national courts, particularly certain French courts, are beginning to recognize ecological damage (as distinct from the devaluation of property) as an independent head of damage (Conseil d'Etat, 28 April 1976, Audibert).

Expenses in respect of emergency and cleaning measures generally qualify for compensation under the general rules on liability, provided that the measures adopted were intended to avoid, limit or remedy specific damage and provided that the effects thereof can be clearly identified. Only Belgium has special rules on this matter, namely Article 7(2) of the Law on Toxic Waste.

It is apparent from the foregoing account that as a general rule the compensation payable for damage is adequate provided that the damage is direct and identifiable and, in particular, that the offender is known.   By contrast, the environment is the poor relation since compensation is rarely paid for the damage suffered by it.   Pollution constitutes damage to the community in general and compensation is therefore not usually payable since no specific interest is directly affected.

3.3.4. The considerations set out in Section 3.3.3. apply only in so far as the person causing the damage is identifiable; if identification is impossible or if problems of causality prevent incontestable proof of the defendant's liability, the victim will generally be denied all compensation.   This problem arises particularly in cases of "fly-tipping" of dangerous waste.

Problems of this type might be resolved by the creation of a compensation fund to be drawn upon in cases where civil law does not provide for reparation and cases where reparation cannot be granted.

Such funds already exist at international level, such as the compensation fund created by the 1971 Brussels Convention for pollution damage caused by hydrocarbons in tanker accidents at sea, and at national level, such as the French guarantee fund (1951) for accidents caused by motor vehicles, which covers physical injuries resulting from traffic accidents on land, and the Netherlands fund for atmospheric pollution.

It is conceivable therefore that a European fund might be introduced for the compensation of victims and the reparation of pollution damage to the environment arising from the transportation of dangerous goods, substances and waste.

3.3.4.1 The person normally entitled to compensation is of course the third party directly affected by the damage. It is therefore necessary to consider what rights of action exist where the victim dies.

Actions of two kinds may be contemplated. The first is an action _ex herede_ whereby the victim's beneficiaries seek compensation for the damage suffered by him. The second is an action for compensation for loss of income.

In studying these rights of action, it is important to note that the concept of "dependants" varies fairly widely from one Member State to another. It ranges from direct ascendants and descendants to collateral relatives (in the United Kingdom) and even, in certain circumstances, cohabitees (Belgium and France and, in cases of mental anguish only, Italy).

3.3.4.2. As regards rights to compensation which may accrue to persons who were on board a vessel or vehicle, such as the driver, helmsman or other crew members, the principles of contractual liability apply together with the national administrative rules concerning the obligations of those engaged in transport operations, and also the international conventions.

As regards employees of transport firms who are the victims of an accident whilst carrying out their duties, they will benefit from the national social legislation on accidents at work and occupational diseases.

3.3.4.3. To sum up, it should be noted that any regulations on liability arising from the transportation of dangerous goods, substances and waste should without fail define the field of application as regards damage for which compensation is payable.

In that connection, it will need to be borne in mind that adequate
compensation is provided in the Member States for cases of death
and injury and also for damage to property where the damage is
direct and identifiable and the person causing it is known.   By
contrast, the present remedies for damage to the environment
(pollution and ecological damage) and damage suffered by the
community as a whole are inadequate.

The problems associated with the recognition of damage specific
to the environment need to be analysed (definition, determination,
problems of long-term exposure, etc).

It would also seem relevant to seek to identify precisely who
should be liable to pay compensation for emergency and cleaning
measures following an accident arising from the transportation
of dangerous goods, substances and waste.

As regards claims for compensation by persons on board a vessel
or a vehicle, it would be appropriate to consider whether damage
suffered by crew members should not, for reasons of equitable
treatment, qualify for compensation similar to that available
to persons unconnected with the transport operation, that is to
say third parties (cf. on this point the views expressed by the
Legal Committee of the IMO in relation to the draft HNS Convention
and those expressed by UNIDROIT in connection with its Draft
Convention on land transport).   As regards damage not qualifying
for compensation under existing ordinary law, consideration must
be given to introducing a compensation fund for damage to third
parties and to the environment.

3.3.5. Limitation of liability

The Member States do not have special rules limiting liability for
damage caused during the transportation of dangerous goods, substances
and waste.   And similarly, limitations of extra-contractual
liability are also rare in the Member States.

There is a case of limitation of liability in the Federal Republic
of Germany, but that applies only to damage inherent in the operation
of the vehicle (StVG Paragraphs 7 and 18).   As a result, damage
caused by the nature of the load carried is not subject to that
limitation.

We should also draw attention to two systems which constitute
exceptions to what has been stated above.   Danish and United
Kingdom courts are sometimes able to limit the compensation awarded
to the victim in the light of the defendant's circumstances.   In
Denmark, compensation for bodily injury and damage to property or to
the environment may be reduced or withheld entirely in cases where
unrestricted application of the liability rules would entail
unreasonably onerous consequences.   In the United Kingdom, the
court may have regard to the financial situation of the defendant
company in determining the compensation to be awarded to the victim.
No such powers are available to the courts in the other Member
States.

In the great majority of the Member States, limitation of liability
exists by virtue of the transposition of certain international
conventions.

International law on road and rail transport does not - other than
with respect to the transportation of nuclear substances - embody
any limitations of extra-contractual liability (for limitations of
contractual liability, reference should be made to the CMR and the
CIM).   In the case of inland waterways, the CLN Convention (1973
Convention on limitation of the liability of owners of inland-
waterway vessels) will limit the liability of the owner of the

vessel; however, because of an insufficient number of ratifications the convention has not yet entered into force.

In maritime law, Article 2(c) of the 1976 LLMC Convention (Convention on the limitation of liability in maritime claims - not yet in force) refers expressly to damage resulting from the infringement of any extra-contractual rights, and the 1969 Convention (International Convention on civil liability for pollution damage resulting from the transportation of hydrocarbons, 1969 and the Protocol thereto, 1976) provides for limitations of liability.

The special nature of maritime conventions lies in the fact that the owner of the vessel must, in order to benefit from the limitation provided for, furnish funds covering the limit of his liability. Once the funds have been furnished, no right to compensation may be enforced against his other property.

However, these limitations of liability do not apply in cases of misrepresentation or personal fault.

Discussions are in progress within the IMO concerning the acceptance, for the purposes of the draft HNS Convention, of the limits of liability for shipowners laid down in the LLMC Convention, the ceilings for which are regarded as too low by some delegations.

Other delegations propose that ceilings should be fixed according to each incident, rather than by reference to the quantities of substances transported.

To sum up, it may be said that all the limitations of civil liability in the conventions referred to are regarded as a concession reciprocal to strict liability, in other words a person who is subject to liability without fault must be in a position to calculate the risks to which he is exposed, particularly in order to enable him to make adequate insurance arrangements.

It is obvious that the question of the limitation of liability
is inextricaly bound up with other problems such as the channelling
of liability (liability attaching to one person alone or to several,
or on a joint and several basis) the possibility of appeals by one
or more persons bearing liability, the problem of insurance cover,
which appears difficult to obtain where liability is unlimited, and
the possible creation of a supplementary fund.

3.4 Attribution of liability

Three main parties are usually involved in transport operations:
the consignor (or the producer, who is generally the person
concerned in the transportation of waste), the carrier and the
consignee (or the party engaged in disposal of the waste).   In
addition to these three parties, there are intermediaries such as
loaders and collectors.   One or more of those parties must therefore
be sought out as under liability to compensate those who suffer
damage in connection with the transportation of dangerous goods,
substances or waste.

We consider below the ways in which this matter is dealt with by
the laws of the Member States and the relevant international
conventions, and also the factors which must be taken into account
in any future Community measure.

3.4.1. National rules on the liability of the producer and the carrier

3.4.1.1. Although there is a clear tendency to deal with the problem
of dangerous waste by introducing specific legislation, so far only
Belgium and France have introduced special rules on liability.

Article 7 of the Belgian Law on Toxic Waste provides that the
producer shall be liable "for any damage whatever which might be
caused by toxic waste, in particular at any time during the
transportation thereof, ..., even if such operations are not
carried out by the producer himself".

This provision not only establishes total civil liability attaching
to the producer for any damage caused by his toxic waste but also
introduces a criterion of causality ("which might be caused ...").
It is therefore the responsibility of the producer to prove the
existence of relieving circumstances such as, for example, the
fault of a third party.

Under French law, the producer is jointly and severally liable
with the carrier, unless he used an officially approved carrier.
In that case, ordinary law applies.

3.4.1.2. The comparable legal rules providing for strict liability
in other Member States relate to the producer only if he is in
possession of the waste (see the WHG in the Federal Republic of
Germany, the rules on "neighbour problems", and the Common Law
approach regarding abnormally dangerous installations). An action
for compensation against the producer becomes inadmissible if the
damage arose after the dangerous waste was placed in the possession
of the carrier.

The majority of the Member States are content to list the obligations
of the consignor, the carrier and the disposal undertaking, but do
not prescribe the specific legal consequences of an infringement.
Any person who fails to fulfil or fulfils unsatisfactorily any
obligation imposed upon him by law will be "liable". Damage caused
to third parties and to the environment falls within the scope of
ordinary law. Under ordinary law, the person liable for the damage
will be the "custodian" of the object in question, in other words
the carrier, on the basis of fault or negligence. These rules
apply without distinction to goods and to waste, except in France
and Belgium.

In exceptional cases, the producer alone is liable where the
carrier has been misled as to the nature or indeed the very existence
of the dangerous waste.

More generally, joint and several liability of the producer and the
carrier is to be expected where the producer is involved in the
act constituting the fault, for example if he requests that the
waste should be discharged into the environment.

Moreover, by virtue of the principles of liability for fault, the victim may institute proceedings for compensation against any party intervening in the transport operation whose fault has given rise to direct damage, namely the loader, the driver's mate, etc.

3.4.1.3. A further problem arises from the fact that it is almost always a company which performs this type of operation. As a result, it is necessary to identify one person to bear the liability amongst the many persons involved.

Although to differing degrees, all the Member States recognize the liability of principals by virtue of acts committed by their agents. The difficulties of enforcing such liability derive from identification of the acts for which the principal is liable. In principle, he is liable only for acts committed by the agent "in the discharge of his duties", a term which is not defined uniformly in the Member States.

In so far as this liability is liability without fault in the majority of the Member States, two specific legislations deserve attention.

The German Civil Code (BGB Paragraph 831) differs from the systems mentioned in that it empowers a principal to prove that he was not guilty of fault in the choice or supervision of his agent. However, court decisions are tending to reduce the scope of this exoneration of the principal.

In Denmark, in the event of damage caused during a transport operation carried out by an employee, the employee and the principal are jointly and severally liable. The Liability for Compensation Act 1984 has now reduced the liability of agents. Under Article 23 thereof, the employer cannot bring an action against an employee unless such an action is deemed reasonable in the light of the fault involved, the situation of the employee and the general circumstances of the case.

3.4.1.4. The same principles must be applied in considering the case where a carrier hires out a vehicle with a driver to a third party in order to carry out a transport operation contracted for. The carrier taking the vehicle on hire is, as far as third parties are concerned, the custodian of that vehicle and is even answerable for any fault on the part of the driver, whose principal he has normally become for the performance of the operation in question.

This general principle is also recognized in contract law by Article 17.3 of the CMR, which provides expressly that the carrier cannot seek to escape liability by enforcing liability for fault on the part of the person from whom he hired the vehicle or agents thereof.

3.4.1.5. A different problem arises if the carrier merely undertakes to cause the transport operation to be carried out and entrusts the performance of that obligation to another carrier.

In that case he acts as a forwarding agent, who undertakes to ensure that goods are transported by entering into the relevant contracts on behalf of the consignor.

In general, he is jointly and severally liable with the carrier chosen, and the position is different only where the carrier subcontracted was designated by the consignor.

3.4.1.6. Each system of rules on liability incorporates "corrective factors" whereby liability is excluded or at least attenuated in certain cases. Such cases of _exoneration_ may be classified as follows:

a. _force majeure_ (act of God), being an unforeseeable and irresistible event such as, for example, natural phenomena; where a court recognizes that a case of _force majeure_ exists, the result is generally total exoneration of the person liable;

b. act of a third party and fault on the part of the victim: the
scope of such cases is viewed in various different ways by the
national courts and is appraised in each individual case
(relative extent of the factors in question); the result is
usually partial exoneration of the person liable or an
apportionment of liability.

In more general terms, it may be said that "wide-ranging" liability,
covering a rather broad area and attaching to several possible
persons, necessitates more cases of exoneration than "narrow" and
clearly defined liability, that is, liability which is limited in
scope and concentrated on only one person.

Thus, the 1960 Nuclear Convention, under which liability attaches
exclusively to the operator of an installation, provides for only
one very limited case of force majeure. By contrast, the Brussels
Convention recognizes force majeure, acts of third parties or
adverse action by a government.

3.4.2. Problems of uniform rules on liability

The different national systems of rules on liability give rise to
considerable uncertainties regarding transportation within the
Community. Moreover, the procedures for enforcing liability are
sometimes unable to guarantee adequate compensation for victims
(problems concerning the solvency of the person liable,
identification of the person liable, etc).

A uniform system of liability specific to the transportation of
dangerous goods, substances and waste, based on strict liability
attaching to one person alone could considerably improve the way
in which victims are treated. Such a system would certainly have
to apply equally to purely national transport operations and
transfrontier operations.

3.4.2.1. Uniform system of liability

If a harmonized system of rules on liability is to be introduced, particular account would need to be taken of the following problems.

3.4.2.1.1. Scope of uniform rules

Article 11(3) of Directive 84/631/EEC concerning <u>transfrontier shipments of dangerous waste</u> requires the Community to draw up rules on liability.   Two main problems arise here:

3.4.2.1.1.1. From the victim's point of view, compensation for damage caused by a transport operation on the basis of rules derogating from national laws regarding transfrontier transport operations is illogical.

3.4.2.1.1.2. From the point of view of potential damage, different compensation and, possibly, different rules on liability for dangerous waste on the one hand and goods and useful substances on the other are inconceivable.

The grave, and sometimes catastrophic, nature of the damage caused by transport accidents is certainly not limited to those involving waste (for example, the accident at Los Alfaques).   Goods and useful substances are no less dangerous than waste in such circumstances.   Moreover, the existence of different sets of rules would encourage "subterfuge".   Rules derogating from the norm for recyclable waste do not therefore seem feasible.

On the other hand, it must be emphasized that the problem of damage caused by the discharge or unauthorized dumping of dangerous waste calls for specific and supplementary rules on dangerous waste (see 3.4.2.1.2.5.).

3.4.2.1.1.3. Harmonized rules in the field in question should be largely <u>multimodal</u> in order to avoid any distortion of competition. It should take account of the work done by UNIDROIT.

As concerns maritime law, the special features of the potential damage involved should be taken into consideration.

3.4.2.1.1.4.   Finally, there is a further problem of a rather technical nature: should a differing approach be taken regarding transportation "in bulk" and "in packages" respectively?

Considerations of this kind are raised in particular within the IMO, where a differing legal approach is adopted inasmuch as transportation in packages is excluded from the draft HNS Convention.

Consequently, that Convention will cover only "all dangerous substances carried in a tank or in a cargo space forming part of the structure of a vessel, without any form of intermediate packaging". The essential basis for this approach is the view that in maritime transport the main risks arise from the transportation of large quantities of dangerous substances in bulk, for example, where a cargo of liquefied natural gas or other hydrocarbons is carried in tankers.

Furthermore, it is also argued that the existing provisions concerning stowage have considerably reduced the risk of accidents.

These considerations do not appear to apply to overland transport. Although a distinction may be drawn in a legal appraisal - in so far as a producer will certainly cease to be "custodian of the structure" where the substance in question is carried in bulk - it would appear unjustifiable to draw a distinction with respect to the risk of accident.

Two arguments in particular persuade us that equivalent legal treatment is called for:

a.   dangerous substances carried in packages can cause equally serious damage (cf. the accident which occurred on 4 September 1984 in the Fourvière tunnel involving blocks of sodium cyanide);

b. stricter conditions as to liability for damage caused by
    transportation "in bulk" might give rise to subterfuges (the
    carrier might be tempted to avoid the stricter conditions as
    to liability).

3.4.2.1.2. Channelling of liability

The transportation of dangerous goods, substances and waste
comprises a series of complex operations in which the carrier,
who is easily identifiable, operates in conjunction with others,
such as the loader, the packer, the consignor, etc, who are much
more difficult for the victim of an accident to identify and
whose roles may vary from country to country.

Here, the channelling of liability would grant the victim access
to a clearly identified interlocutor against whom he could enforce
his rights, independently of the fault of the person liable.

The basis for this legal device is strict liability, where the
concept of "risk", or to be more precise the apportionment of risks,
is of primary importance.   It is even conceivable that the sharing
of risks could extend beyond the persons directly involved in the
transport operation.

Additionally, the channelling of liability in the strict sense
implies that a victim can seek a remedy only against the person
identified by law as the focus of liability.

3.4.2.1.2.1. Before the matter is studied in greater detail, it
is necessary to define the terms producer, consignor and carrier.

Maritime law relies upon the concepts of "shipper" and "shipowner", but difficulties exist regarding the efinition of a "shipper". The most recent proposals of the IMO Legal Committee define the shipper as "a person who has decided to cause dangerous substances to be transported by sea".

Transposed to overland transportation, the concept of shipper or consignor may be likened to that of "producer" as hereinafter used, without there being any need to think in terms of ownership of the goods. In fact, ownership of the goods may very well pass to the consignee when they are entrusted to the carrier and, furthermore, the "owner" of waste may also relieve himself of liability by abandoning the waste (derelictio).

In the case of transportation by inland waterway, the term "owner of the vessel" seems more appropriate to define the person liable for the transport operation since the considerations of maritime law leading to its adoption appear equally applicable to this type of transport.

As far as overland transportation is concerned it will be necessary, in defining the carrier, to refer to the party to the transport contract, regardless of whether he actually owns the vehicle (cf. cases where the vehicle is on hire).

3.4.2.1.2.2. There is a "concentration" of liability under the Convention on civil liability in the field of nuclear energy (Paris 1960), which attaches all liability to the operator of the nuclear installation in the event of an accident during transportation.

The Brussels Convention (International Convention on civil liability for damage due to pollution by hydrocarbons, 1969) also attaches all liability in the event of an accident to a single person: the shipowner. However, Article III(4) thereof expressly excludes any action for compensation against the servants or agents of the owner, but not other actions. It must therefore be concluded that actions against the other parties involved continue to be available.

Recent thinking on the HNS Convention (Draft Convention on
liability and compensation in relation to the transportation of
toxic and dangerous substances by sea) shows a tendency to restrict
the number of persons who may be liable.   The IMO Legal Committee
decided at the end of 1984 to propose a system of dual exclusive
liability whereby the shipowner would be liable up to a certain
limit and the shipper of the substance would be liable for damage
in excess of that limit.

Another example of concentration of liability is to be found in
the present draft UNIDROIT Convention (draft articles of a
convention on civil liability for damage caused during the
transportation of dangerous goods by road, rail and inland waterway
- 1985 version).   According to that draft, the carrier is liable
for any damage caused during transportation since it is he who has
control and custody of the products.   This liability is without
prejudice to any recourse he may have against third parties.
However, the draft contains proposals for two important exceptions
(cf. variations I and II):

a.   total exemption of the carrier from liability where loading
      operations are carried out under the sole control of a third
      party;
b.   joint and several liability where loading is carried out by
      someone other than the carrier.

Article 7 of the UNIDROIT draft provides for concentration of
liability in that it expressly excludes recourse against certain
persons such as, for example, the carrier's servants, the owner,
hirer or operator of the vehicle, the pilot or driver, etc.

The draft United National Environment Programme (Draft guidelines
concerning the management of dangerous waste) also adopts the
principle of the channelling liability.   Under that draft, each
State must specify a natural or legal person who will bear
liability for any damage resulting from transportation, storage,
handling or disposal of dangerous waste.   As a rule, that person
should be the producer of the waste.

Thus, only under the 1960 Nuclear Convention is liability <u>concentrated</u> <u>exclusively</u> on one person.   It has, of course, to be borne in mind that the nuclear field has its own specific characteristics - the number of parties involved is small and, in most cases, they are of a public or semi-public nature, so that it may be said that in reality State liability is involved.   However, in view of the very numerous transactions and parties involved in the transport operations under review here, we cannot take the nuclear convention as a model.   Furthermore, it would seem unfair to attach sole liability to the producer (in comparison with the operator of a nuclear installation) and at the same time to limit his right of recourse against others.   In such circumstances, the carrier's fulfilment of his obligations of care can be guaranteed only by the introduction of a system of criminal-law and administrative penalties; it is not a matter which falls within the powers of the Community.

3.4.2.1.2.3. All the discussions on the channelling of liability for damage occurring during transportation relate essentially to the carrier and the producer (manufacturer).   The essential arguments put forward may be summarized as follows:

<u>For liability attaching</u>

to the carrier	to the producer
	- "polluter pays" principle - the agent causing the pollution (production) must pay for its consequences
- physical, intellectual and legal rights over the load - control at the time of the accident	- the carrier is the custodian of the behaviour of the object, but has no power over its structure; the producer is the custodian of the structure
- acceptance of risks by taking part in a commercial activity	- better distribution of financial risks, particularly in cases of catastrophic damage

in the case of railways, State
companies or the State itself
can bear the financial risk

- preventive effect to avoid
  damage

- preventive effect to avoid
  production of dangerous waste
  and, in some cases, dangerous
  substances

- identification easy for the
  victim

- liability cannot terminate when
  the load is entrusted to a
  carrier; transport is only one
  element of marketing (goods and
  substances) or disposal (waste)

- high level of insurance
  premiums might provoke
  concentration in the transport
  sector: not only quantitative
  concentration but also
  qualitative concentration,
  which is in fact desirable.

- adequate insurance cover could
  be more easily purchased by the
  producer in view of the high
  premiums

- the producer's liability would
  be merely the counterpart of
  his liability for defective
  products.

3.4.2.1.2.4. The problem of the channelling of liability is closely
related to the problem of insurance cover. Attaching liability to
a specified person makes no sense unless the person liable can
guarantee full compensation for the victim.

Prima facie, such a guarantee could be obtained by insurance cover
which extended to the high levels of potential damage, against
payment of premiums which the insured could afford. Thus, in the
interests of both the victims and those liable, compulsory insurance
should be introduced throughout the Community both for the purely
national and for the transfrontier transportation of dangerous goods,
substances and waste, as has already been done in Belgium and Germany.

The problems of insurance law involved in the introduction of such
a measure, such as scope of application, kinds of damage to be
covered, criteria of insurability, etc are, of course, closely bound
up with the answer to the problem of the channelling of liability.
Their examination falls outside the scope of this study.   In view
of the scant statistical information available, specific research
would certainly be required.   These remarks apply also to the
introduction of a system of financial guarantees, deposited either
by those themselves involved in transport operations (1) or by the
State.

3.4.2.1.2.5. The arguments so far put forward take little account,
for the purpose of effectively channelling liability, of the
specific problem connected with the transportation of dangerous
waste, namely unauthorized dumping or tipping.   As has been shown
by the "affair of the Seveso barrels", damage or the threat of
damage may come about or be detectable only after a fairly long
period of time.   In fact, the problem is not so much a transportation issue as a special issue of waste management.

In such circumstances the only person to whom liability could
possibly be channelled would be the producer and account will need
to be taken of this particular aspect in introducing any harmonized
system of liability.

_____

(1) Readers are referred to the account by M. H. Smets of the OECD,
     who argues in favour of the introduction of such a system, in
     his article "Le système de la garantie pour mieux contrôler le
     mouvement transfrontalier de déchets dangereux (The guarantee
     system for better control of the transfrontier movement of
     dangerous waste)", in "Transports transfrontières de déchets
     dangereux (Transfrontier shipments of dangerous waste)",
     OECD 1985.

3.4.2.1.2.6. Before putting forward alternative solutions, based on the foregoing considerations, it would appear relevant to refer briefly to Article 15 of the Commission's proposal on transfrontier shipments of dangerous waste, as originally submitted.

That Article, which was subsequently withdrawn, provided for <u>strict liability</u> in respect of waste attaching exclusively to its producer "from the time of production of the waste until disposal thereof". Thus, the producer would have been "civilly liable for any damage caused to any third party by his waste until safe disposal thereof".

Article 15 did not provide for rights of recourse against third parties at fault, or even clauses covering exemption or limitations of liability.

In the light of the foregoing considerations, the weaknesses of such a system, which is even more rigid than the provisions on liability in the convention governing the transportation of nuclear substances, are obvious. They may be summarized as follows:

a. in the absence of corresponding criminal-law or administrative penalties applicable to the carrier or any other party involved, the exclusion of rights of recourse is unjustified;

b. the omission of exemption clauses would even go beyond the already very inflexible model of the nuclear convention;

c. without any limitation of liability, insurance cover would be difficult to obtain.

It appears that there are three alternative solutions.

1. <u>Strict liability of the producer accompanied by the possibility of rights of recourse</u> in the event of wrongful conduct on the part of the carrier or other persons. In addition, there would need to be <u>exemptions from liability</u> in cases of <u>force majeure</u> and acts of third parties, and also a <u>limitation of liability</u>, the ceiling for which might vary according to whether a compensation fund was created or not.

Such a fund could be financed by contributions from the producers of dangerous substances and waste. The benefit of the limitation of liability would be withheld in the event of fault. A system of compulsory insurance would also be required, or some other form of financial guarantee.

2. Strict liability of the carrier with the possibility of rights of recourse against the consignor (if, for example, the latter failed to give notification of the dangerous nature of the load) or against any other third party at fault. Exemptions from liability in cases of force majeure or acts of third parties would need to be provided for, and likewise a limitation of liability. The extent to which liability was limited could be based on the number of vehicles at the disposal of the carrier. Needless to say, a system of compulsory insurance is required.

3. A system of "mixed liability" of the type at present suggested within the IMO seems less desirable in the context of overland transport. In fact, if that system were transposed to overland transport, it would imply:

a. liability of the carrier up to a certain limit;
b. additional liability of the producer, likewise limited;
c. liability of the producer to third parties in cases where the carrier is financially unable fully to meet his obligations.

The disadvantages would lie above all in the fact that in such circumstances two persons would have to provide financial guarantees, which means duplicated insurance and therefore the double payment of premiums, giving rise to increased transport costs. Moreover, it would be difficult to define the threshold of the producer's liability, and at the same time the fixing of the insurance premiums would be a delicate matter. Added to this is the fact that, in adopting this solution, the IMO is referring to an existing convention which governs limitations of liability, whereas in the field of overland transport no such international regulations exist.

3.5. Criminal liability

By contrast with liability under civil law, liability under criminal
law can arise only if a law is infringed.   Moreover, under criminal
law the penalty is imposed specifically in relation to the wrongful
act (note, however, the liability attaching to employers under French
criminal law on the environment), and it follows that liability will
inevitably be dispersed.

The penalties imposed vary greatly from one Member State to another
depending upon how the offence is classified.   Thus, in certain
States the least serious infringements are governed by administrative
law rather than by criminal law, whereas in other countries there
is a trend towards the "criminalization" of acts involving danger
or damage to the environment.

Three types of criminal rules may apply depending upon the terms in
which the charge is made:

a.   infringements of traffic and transportation regulations, and
      more particularly infringements of the national counterparts
      of international conventions such as the ADR, the RID, the
      ADNR and the IMDG Code;
b.   infringements of environmental law, such as, for example, bans
      on the discharge, wilful or otherwise, of certain substances
      or waste into the environment;
c.   infringements of general regulations designed to protect the
      physical integrity of persons and public order and safety.

All the Member States impose penalties for infringements of the
regulations on the transportation of dangerous goods, substances
and waste, but they vary considerably from one country to another
and according to the consequences of the infringement.

Whereas France and Germany impose heavy penalties (for example, in
the Federal Republic of Germany under Paragraph 330 of the Penal Code,
the penalties for serious cases extend from 6 months' to 10 years'
imprisonment), in other countries, such as Denmark and Ireland, such
penalties are imposed only on the basis of more general charges.

For instance, the transport legislation in such countries provides
only for fines which in certain cases are ridiculously low because
they have not been updated (the example of Ireland speaks for itself,
since an infringement of the Transport Act 1950 entails a maximum
fine of £Ir 10).

The same situation prevails in purely environmental matters, where
the sentence of imprisonment ranges from a minimum of 8 days
(Belgium) to a maximum of 10 years (Federal Republic of Germany).

The criminal legislation in the Member States thus differs very
greatly, which may raise problems in more than one respect.   It
is clear that the lesser degree of severity in certain States
could lead to greater recourse to subterfuge.

Likewise, certain possible options regarding the channelling of
civil liability may depend upon the applicable criminal-law rules.
For instance, if all civil liability were attached exclusively to
the producer, it would be necessary to offset this by introducing
corresponding criminal-law penalties to ensure that carriers
observed their duty of care.

4. FINAL CONCLUSIONS

4.1. We shall start by giving a summary of the main conclusions of each of the national reports.

The Belgian report concludes that there are three major types of problem:

A. Clarification of the division of powers

The transportation of dangerous goods, substances and waste falls simultaneously within the powers of several central administrations, regional administrations and various local authorities.

B. Gaps in the legislation

Not only does it appear that Belgian law has not yet incorporated a series of European directives but also that there are gaps in the legal provisions concerning:

a. the notion of "recyclable waste";
b. standard limits of concentration for dangerous waste;
c. the status of the various ancillary operators involved in transport;
d. the accompanying documentation for transport operations;
e. controls on international movements of waste.

C. The definition of a coherent policy strategy

In Belgium, the problems of dangerous transport operations have not been dealt with in conjunction with an overall policy for the prevention of risks and/or the rational management of waste.

The Danish report focuses upon certain legal difficulties of interpretation of the rules applicable to the transportation of dangerous goods and the transportation of dangerous waste.

It emphasizes that, whilst certain divergences in the applicable legal regimes are justifiable by virtue of the mode of transport involved, that is not always the case. Consequently, it identifies the need for greater international co-operation regarding safety regulations and specific rules governing civil liability and insurance.

The French report ends with a very critical appraisal of the legal regime in force; it points to the increase in the number of provisions of a technical nature, lacking any order of priority or structural arrangement, as a major failing. Old rules are overlaid with modern requirements which vary according to the date on which the means of transport or equipment in question was put into service: as a result, there is a wide range of transitional and permanent regimes which is most damaging to the clarity, and therefore to the effectiveness, of the system.

This disadvantage is exacerbated by the fact that the rules in question are mainly of a markedly technical nature and sometimes display impenetrable legal complexities.

In addition, the maze of regulations is strikingly devoid of any coherent thread or logical framework to serve as a guide for interpretation. A major effort must therefore be made in the years to come to clarify and interrelate the provisions concerned.

The Irish report refers to the need for modernization of the legal provisions now applicable to the transportation of dangerous goods, substances and waste and emphasizes the need to ratify the ADR.

The Italian report draws attention to the recent promulgation of a whole series of specific provisions concerning the management of dangerous waste and states that it is therefore too early to evaluate the effectiveness of those rules.

As regards the transportation of dangerous goods, on the other hand, it emphasizes the existence of very old rules whose effectiveness can indeed be assessed. Here, although the legal conditions applicable to rail transport are satisfactory the same is not true of the rules on road and sea transport.

The carriage of dangerous goods by road suffers from a multiplicity of sources of applicable law; the rules are fragmentary and unco-ordinated, not to say inadequate.

For the transportation of dangerous goods by sea, there is no uniformity between Italian legislation and that of the other States: greater efforts directed towards legislative harmonization are needed at international level.

Special consideration should be given to:

a.   a classification of dangerous goods which applies regardless of the mode of transport;
b.   uniform specifications for containers, regardless of the mode of transport.

The Luxembourg report arrives at conclusions similar to those of the Belgian report, except that apparently the European directives on waste have already been transposed into Luxembourg law.

The Netherlands report clearly shows that two types of legislation coexist in this field; legislation on environmental protection and legislation on transport safety.

It points out that the existing international conventions do not devote sufficient attention to environmental objectives. It calls for a European policy of active harmonization of all rules directed towards safety in transport and protection of the environment. Moreover, it stresses the need to:

a. establish a precise legal status for dangerous waste relating to that of dangerous goods;

b. guarantee compensation for damage caused by dangerous goods, substances and waste by introducing compulsory insurance and a system of strict civil liability;

c. ensure compliance with and enforcement of the rules in force by means of stricter measures;

d. integrate the rules on transport accidents with the legislation on environmental protection, and integrate the notification rules applicable to environmental protection with the legislation on transport.

The German report concludes that it is necessary and urgent to render the rules uniform and to simplify them.

It stresses the inadequacy of the rules on training for drivers and the rules restricting transport routes.

It calls for a policy favouring the carriage of goods by rail. As regards exports of dangerous waste, it points out various gaps in the existing system of control under German law and Community law. It notes a trend in German legislation towards the principle that waste should be disposed of in the country of production: that principle would, however, be contrary to Community law. As regards liability, the German report states that the rules creating the legal basis for entitlement to compensation should be simplified: fundamentally, priority should be given to a system of liability without fault covering pretium doloris. The system of compulsory liability insurance should be harmonized and extended; and consideration should be given to the possibility of using State-controlled insurance pools where private companies' capacity to provide cover is exceeded.

The United Kingdom concludes that there has been proper transposition into national law of the existing international provisions and that the problem of gaps in the legal provisions is essentially confined to the sector of dangerous waste.

However, it stresses that effective application of the law continues
to be a major problem in this area.

"Losses" of waste into the natural environment are particularly to
be feared.   In view of the highly technical nature of the many
rules to be observed (and in respect of which control is required),
the publication and publicizing of codes of good practice are
recommended.

4.2. General conclusions

It is of course very difficult to make a summary analysis of this subject. Its main characteristic is without doubt its complexity. The principal reasons for this complexity derive from the fact that the matters under review:

a. touch on two areas of the law which might be regarded as separate but which in fact overlap extensively, namely transport law and environmental law;

b. involve an intricate mass of rules of which the sources and scope differ greatly and which, moreover, belong to different legal systems;

c. are constantly being amended and updated;

d. are essentially highly technical;

e. lie at the heart of disputes concerning the extent of Community powers.

In the latter respect, it must be strongly emphasized once again that the Community does in fact have full powers regarding both transport and waste management. To be more precise, it may be said that the Community has exclusive powers in those areas and that those powers have not yet been fully exercised either internally or internationally.

4.2.1. An analysis of the relationships between the individual legal systems involved highlights the very specific nature of Community law.

At the present time there is a great need for harmonization of the rules applicable within the European Community to the transportation of dangerous goods, substances and waste: the task of legal harmonization is fully consistent with the rightful role of the EEC.

Accordingly, it is particularly appropriate to adjust or readjust the requirements of Directive 84/631/EEC on transfrontier shipments of dangerous waste.

The logic of creating a large internal market, which is the essential logic of the EEC, involves the abolition of any obstacle to the free movement of goods and services and, in particular, the abolition of national frontiers.

Since dangerous waste is classifiable as goods, it should benefit from the conditions on freedom of movement; that does not mean that there should be no Community rules specifically applicable to waste management - what it does mean is that we must not confine our thinking merely to "the crossing of frontiers".

In other words, total harmonization is now called for, together with the obligation to give notification of movements of waste, whether such movements are internal or international.

Likewise, it is to be hoped that in the future the Community will succeed in speaking with one voice at international level, and indeed this must certainly be achieved if we wish to avoid jeopardizing the progress made or to be made internally.

4.2.2. The existing law and improvement thereof

A review of international law on the transportation of dangerous goods, substances and waste shows that there are numerous rules requiring harmonization.

Genuine Community law must be created in this area. It must be inspired by the existing international provisions and recommendations. As far as possible, it must lay down common requirements irrespective of the mode of transport. Above all, it must prescribe the same rules for national and international transport operations.

In particular, it would appear that binding Community action is called for in order to ensure effective implementation of a whole series of provisions concerning maritime transport.

The distinction between "products" and "waste" does not seem to be of fundamental importance with respect to the Community law to be created regarding dangerous transport operations. However, that does not mean that specific Community rules on dangerous waste are not justified as an additional requirement. The very nature of waste means that it raises a specific problem relating to transport: deliberate unauthorized dumping.

The existing Community rules on dangerous waste must therefore be retained, with amendment in certain respects. They must be extended to both national and international transport operations and must be supplemented by a valid common definition of "dangerous waste".

The list of national regulations on the transportation of dangerous goods, substances and waste confirms the need for Community harmonization in this area. There are numerous enactments, drawn up at different times, which render the whole subject extremely complex (and even contradictory). It goes without saying that this situation detracts from legal certainty and just compensation for any damage likely to be caused.

A major gap in certain national legislation derives from the absence of a clear definition of what constitutes waste and the particularly difficult status of "recyclable waste". Community harmonization should deal with these problems.

In view of the fact that several national legislatures have adopted definitions of "dangerous waste" incorporating quantitative standards, it is clear that the EEC should proceed with harmonization to the point of establishing a common list of waste which includes quantitative standards.

As regards the persons upon whom obligations are imposed, scrutiny of the national laws also reveals a degree of diversity which calls for EEC action to harmonize and simplify the rules.

A specific legal status should be laid down for the various intermediaries involved, in particular collectors of dangerous waste.

It is of absolutely primary importance to systematize at Community level the obligations incumbent upon the principal operators involved in transporting dangerous goods, substances and waste. This will provide the best guarantee of greater legal certainty and an appropriate solution to numerous problems of civil liability. In fact, if the obligations of consignors and carriers are clearly formulated, it will often be possible to attribute civil liability to one or the other on the basis of a specified fault or the materialization of a specified risk.

As regards the carrier's obligations, it is self-evident that special training for dangerous goods and waste should be accorded priority. This matter should be the subject not only of EEC rules but also of incentive and information schemes.

The obligations of the various persons involved in the loading and unloading of goods and waste certainly merit special attention on the part of the Community legislature.

In addition, an important step to be taken at Community level is incontestably the adoption of uniform emergency cards to accompany goods and waste.

With respect to accidents, the existing databanks should be amalgamated with a view to promoting greater safety.

In any event, harmonization and simplification are the two major areas in which Community intervention is appropriate.

4.2.3. Liability regarding the transportation of dangerous goods, substances and waste

The rules on extracontractual liability in transportation normally have a twofold purpose:

a. to provide compensation for damage suffered by third parties as a result of transport accidents;
b. to prevent future damage.

Account should be taken also of damage to the environment and the ecological balance, even if only partial reparation is so far available for such damage. The national reports and the consolidated report show that these areas of positive law suffer from gaps which are in part due to the differing rules on liability.

The main preoccupation of a third party who has suffered damage as a result of an accident is to obtain compensation. It is therefore of no interest to him whether the carrier, producer or any other person involved in the transport operation caused the accident through "fault" or not.

However, the majority of Member States of the European Community have a system of liability based on the concept of fault, and moreover such fault must be proved by the victim. What is more, the victim is also confronted with the problems of causality, in relation to which the burden of proof also falls upon him to a considerable extent.

Starting from this unsatisfactory position, consideration of the problems raised leads us to the following conclusions regarding the introduction of a unified (harmonized) system of liability and compensation for damage caused during the transportation of dangerous goods, substances and waste:

The main purpose of such unified regulations must be to ensure
adequate compensation for victims

This objective must be attained by taking account of the following
points:

1. Such a system of liability should, in conformity with the
    international conventions in existence or in course of
    preparation, be based on strict liability which reverses
    the burden of proof in order to improve the victim's
    situation.

2. The scope of such a system should not be limited to
    transfrontier transport operations. From the victim's
    point of view, differing regulations for purely national
    transport operations are illogical.

3. Such regulations should take account of the fact that this
    matter is above all a question of transport and that
    accordingly it is inappropriate to differentiate between
    goods, substances and waste. To treat "waste" as a separate
    item is meaningless except with respect to unauthorized
    dumping, which must be taken for what it really is, namely,
    essentially a problem of evasion of the law.

4. As regards the definition of what is "dangerous", the
    suggestions put forward by UNIDROIT should be taken into
    account, namely:

    a. a definition of dangerous substances;
    b. a special list;
    c. restrictions on the transportation of certain
        exceptionally dangerous substances.

5. <u>Such regulations should be multimodal</u> and should make exceptions only for sea transport, for which the existing rules, and likewise the specific features thereof, cannot be ignored (subject to the need for specific rules to be formulated concerning the management of dangerous waste to be disposed of at sea).

6. Such regulations <u>should not draw a distinction between the legal treatment for transport operations "in bulk" or "in packages"</u>. As we have seen, both types of transport operation involve the risk of accidents.

7. <u>Channelling of liability</u> on to a person should take account of the following considerations:

   a. it would not appear politically possible to impose <u>exclusive liability</u> on the same basis as in nuclear matters;

   b. <u>rights of recourse</u> against third parties guilty of fault ought to be available;

   c. <u>cases of exemption</u> from liability should be clearly defined, such as <u>force majeure</u> and acts of third parties;

   d. a <u>limitation of liability</u> is undoubtedly called for, with due regard for the economic needs and rights of the victims.

8. As regard the <u>damage to be covered</u>, <u>damage to the environment</u> (pollution and ecological damage) should be defined and, like <u>emergency and cleaning measures</u>, should be incorporated in the system of liability.

9. With respect to <u>damage for which compensation is unavailable</u> (for example because of non-identification of the person liable or owing to problems of causality), the creation of an <u>indemnity fund</u> should be considered.

10. As regards persons suffering damage, consideration should
be given to including within the category of beneficiaries
of the rules on liability the persons on board the vessel
or vehicle (crew members).

11. Adequate compensation of victims must be guaranteed:

a. this objective should be attained above all by
introducing a system of compulsory insurance for the
transportation of dangerous goods, substances and
waste;

b. in addition, the possibility of requiring a
financial security to be deposited must be considered,
with forfeiture of that security possibly also viewed
as a penalty.

LIST OF ANNEXES

1. List of international conventions, recommendations and draft instruments concerning the transportation of dangerous goods, substances and waste

2. Financial cover for damage arising from the transportation of dangerous waste

3. List of national reports

4. Working plan and questionnaires for preparation of the national reports

ANNEX 1

LIST OF INTERNATIONAL CONVENTIONS, RECOMMENDATIONS AND DRAFT
INSTRUMENTS CONCERNING THE TRANSPORTATION OF DANGEROUS GOODS,
SUBSTANCES AND WASTE

1. Transport and safety

- Recommendations drawn up by the United National Committee of
  Experts on the transportation of dangerous goods (1984 edition)

- International Regulations concerning the Carriage of Goods by
  Rail (RID)

- European Agreement concerning the International Carriage of
  Dangerous Goods by Road (ADR)

- European Recommendations concerning the International Carriage
  of Dangerous Goods by Inland Waterway (ADN)

- European Agreement concerning the Carriage of Dangerous Goods on
  the Rhine (ADNR)

- International Maritime Dangerous Goods Code (IMDG Code)

- Emergency Instructions for Vessels carrying Dangerous Goods -
  Emergency Cards (1985 edition)

- Guide on Emergency Medical Treatment to be given in the event
  of Accidents involving Dangerous Goods (GSMU, 1985 edition)

- Recommendations on Safety in the Transportation, Handling and
  Storage of Dangerous Substances in Port Areas (1985 edition)

- Code of Safe Practice for Solid Bulk Cargoes (1983 edition)

- International Code for the Construction and Equipment of Ships
  carrying Dangerous Chemicals in Bulk (IBC Code)

- International Code for the Construction and Equipment of Ships
  carrying Liquefied Gases in Bulk (IGC Code)

2. Transport in general

- International Convention concerning the Carriage of Goods by Rail (CIM, 1970)

- Convention concerning Contracts for the International Carriage of Goods by Road (CMR, 1956 and 1978 Protocol)

- Convention concerning International Rail Transport (COTIF, 1980)

- International Convention on the Unification of certain Rules on Bills of Lading (1924 and 1957 Protocol)

- United Nations Convention on the Carriage of Goods by Sea (1978, not yet in force)

- United Nations Convention on Multimodal International Transport (1980, not yet in force)

- International Convention for the Safety of Life at Sea (SOLAS 1974 and 1983 Amendments)

3. Pollution and waste

- International Convention on the Prevention of Pollution by Ships (MARPOL 1973£78)

- Oslo Convention of 15 February 1972 for the Prevention of Marine Pollution by Dumping from Ships and Aircraft

- London Convention of 29 December 1972 on the Prevention of Marine Pollution from the Dumping of Waste

- London Convention of 2 November 1973 for the Prevention of Pollution by Ships (amended by the Protocol of 17 February 1978)

- Bonn Agreement of 9 June 1969 on Co-operation to Prevent Pollution of the North Sea by Hydrocarbons

- United Nations Environment Programme Draft Convention on Toxic and Dangerous Waste

- Decision and Recommendation of the OECD Council on Transfrontier Movements of Dangerous Waste (1984)

4. Liability

- Convention concerning Limitation of the Liability of Owners of Inland Waterway Vessels (CLN, 1973 and 1978 Protocol)

- International Convention on Civil Liability for Damage due to Pollution by Hydrocarbons (CLC, 1969 and 1976 Protocol)

- 1976 Convention on the Limitation of Liability in Maritime Claims

- International Convention establishing an International Indemnity Fund for Damage due to Pollution by Hydrocarbons (1971)

- Draft Convention on Liability concerning the Carriage of Toxic and Dangerous Substances by Sea (HNS)

- UNIDROIT Draft Convention on Civil Liability for Damage caused during the Carriage of Dangerous Goods by Road, Rail and Inland Waterway (March 1985 edition)

5. Community Directives concerning Waste

- Council Directive 75/442 on waste

- Council Directive 75/439 on the disposal of waste oils

- Council Directive 76/403 on the disposal of PCBs/PCTs

- Council Directive 78/319 on toxic and dangerous waste

- Council Directive 84/631 on the supervision and control of transfrontier shipments of dangerous waste

- Proposal for a Directive amending Directive 84/631/EEC (Notification of non-member countries)

Functioning of the international organizations in this field

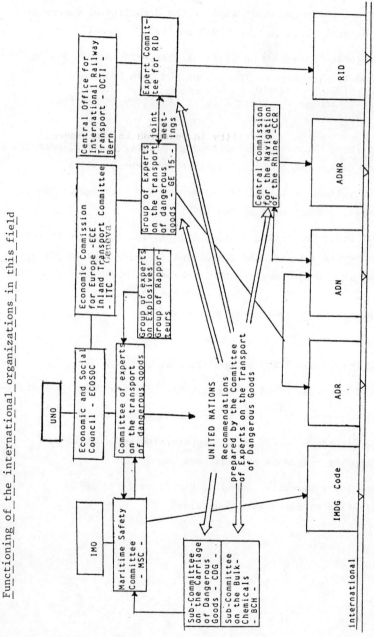

ANNEX 2

FINANCIAL COVER FOR DAMAGE ARISING FROM THE TRANSPORTATION OF
DANGEROUS WASTE

Outline of the problem

Any Community rules on liability in relation to the transportation
of dangerous waste must be designed to provide adequate compensation
for any damage caused during such transportation to persons,
property and the environment.

This aim can be achieved only partially by channelling legal
liability onto one of the persons involved in the transport
operation.   Cover for such damage must be guaranteed.   Guaranteed
cover can be secured only by three means which, in our opinion,
must be studied and analysed as a single unit:

a.   the introduction of compulsory insurance for the transportation
      of dangerous waste;
b.   the deposition of a financial security (such as that suggested
      by the OECD, which would be furnished by the carrier and
      forfeited in the event of an accident);
c.   the creation of an indemnity fund for damage not covered
      under the liability rules or insurance.

It was in the light of those considerations that the Council of
Ministers of the European Communities called upon the Commission,
in Article 11(3) of Directive 84/631/EEC, to prepare by September
1988 proposals not only on liability but also on a system of
insurance relating more particularly to transfrontier shipments of
dangerous waste.

I. Introduction of a system of compulsory insurance for the transportation of dangerous waste

Except in two Member States, there is no compulsory insurance specifically for national and transfrontier shipments of dangerous waste. This is due to the fact that, quite apart from purely economic factors (for example, the number of firms in this sector is limited), the following matters are regulated differently from one Member State to the other and the introduction of a harmonized system of compulsory insurance would presuppose a harmonized approach to them:

1. The channelling of liability for damage arising from the transportation of dangerous waste

Problems: What are the consequences, in terms of insurance law, of channelling liability (as regards the producer, the carrier and the disposer)? What economic considerations are to be taken into account in this context? Concentration of liability upon the carrier may, for example, lead to distortions of competition since small firms might find it difficult to effect onerous insurance policies (catastrophic damage). Concentration of liability upon the producer raises, among other things, the problem whether the insurance market would be prepared to cover risks of accident where the producer is not covered in the same way as the carrier.

2. What types of damage can be covered?

This problem arises in particular with respect to indirect damage to property and ecological damage. Could American law, which takes matters further, be taken as a model?

3. Extent of cover for damage

Problems: a particular feature of damage caused by dangerous waste is that it occurs rarely and is very expensive (catastrophic damage in certain cases). This gives rise to great difficulties in calculating the insurance premiums.

In particular, a study should be undertaken of the limitations and ceilings of insurance compensation existing nationally, and their effect on any Community action in that regard.

The legislation on limitation of liability for road and rail transport varies from State to State. There is no limitation of liability (except regarding nuclear materials) for damage that may be caused during the transportation of dangerous substances.

Where there is no legal ceiling on liability, it may be difficult to obtain insurance cover.

There are other questions to be considered in this context:

a. What factors influence the extent of insurance cover where there is no compulsory insurance?

b. To what extent must such factors be taken into account for the introduction of a compulsory insurance system (for example specific characteristics of the waste to be transported)?

c. Financial capabilities of the insurance market.

d. Possibilities of special agreements providing for the creation of insurance, reinsurance and joint insurance "pools".

4. Definition of the types of waste to be covered by compulsory insurance

There are three possibilities:

a. uniform definition of dangerous waste;

b. preparation of a special list;

c. limitation to certain particularly dangerous types of waste.

5. Which one or more of the factors set out below should render
    the insurance cover operative in the event of compulsory
    insurance being introduced?

a. Behaviour giving rise to damage?
b. Occurrence of damage?
c. Application for compensation?

Problem: all three courses of action are followed in different
ways on the European insurance market.

6. Rules on multiple claims and series of claims

How must such claims (that is to say, claims which relate to the
same circumstances or the same cause of damage) be treated,
particularly as regards transfrontier transport?

7. Criteria of insurability

Problems: These criteria such as fortuitousness, suddenness and
lack of intent are dealt with differently in the Member States.
To what extent must these customary criteria be reviewed in relation
to environmental damage? Another problem derives from the varying
rules on exemption from liability.

8. Economic and political considerations connected with
    compulsory insurance

For example:

a. To what extent can the insurance premiums be borne by small
    firms?
b. To what extent (with a system of liability based on risk) is
    the insurance market capable of covering the risks (especially
    for extensive damage)? To what extent are insurance
    statistics available on the transportation of dangerous waste?

c. What obstacles stand in the way of the effective application
    of such a system of compulsory insurance (comparison with
    the difficulties encountered when compulsory motor-vehicle
    insurance was introduced in certain Member States)?

d. Is it possible to control such a system of compulsory insurance,
    as for example in the case of transportation of nuclear
    materials?

e. How can cover be provided for damage caused by uninsured or
    unidentifiable vehicles (question of indemnity funds)?

II. Deposition of a financial security for the transportation of dangerous waste

The deposition of a financial security by those engaged in the
transportation of dangerous waste (producers or carriers) may be an
effective instrument, particularly for the avoidance of environmental
damage and control of transfrontier transport. The forfeiture of
such securities by way of penalty (see OECD proposal) might enhance
safety in such transport operations.

In several European States comparable systems already exist which
are designed to prevent environmental damage caused by waste (for
example, Sweden and Norway require a financial security to be
deposited when a new vehicle is purchased; the security is repaid
when the old vehicle is returned). The legal, economic and political
feasibility of such a measure (in parallel with - or possibly instead
of - certain measures in the field of liability and insurance) should
be considered.

Questions:

a. who should be obliged to deposit the security?

b. What should be the amount of the security?

c. Precise conditions for forfeiture of the security?

d. Rights of recourse of the person depositing the security?

e. Possibilities of control (inspection upon dispatch and arrival)?

III. Creation of an indemnity fund

Claims in respect of accidents involving dangerous waste encounter obstacles above all in two cases:

a.   in view of the personal nature of the right to compensation, there is no reparation for damage to the general community;
b.   entitlement to compensation arising from damage to individuals is not enforced because of legal obstacles, such as those relating to causality.

The creation of an indemnity fund might resolve these problems. Its purpose would be to ensure compensation for persons whose rights thereto cannot be enforced for reasons relating to the laws on liability or insurance.

For this, it is particularly important to distinguish the following areas of application:

a.   damage for which no right to compensation exists under the civil law on liability;
b.   damage caused by persons who cannot be identified;
c.   ecological damage for which compensation is unavailable or only partly available by reason of the personal nature of the right to compensation.

It would be useful in this context to make a comparative study of existing indemnity funds, such as the Netherlands fund for air pollution compensation (likewise the Maine Coastal Protection Fund in the USA and the Japanese compensation fund for damage to health).

Questions arising:

a.   procedural questions such as proof of the non-enforceability of rights, subsidiarity of recourse;
b.   recourse by the fund against the polluter;
c.   extent of compensation to be paid;
d.   financing of the fund (for example, contributions from the industry concerned).

ANNEX 3

LIST OF NATIONAL REPORTS

Belgium and Luxembourg J.P. Hannequart, Geer, Belgium

Denmark Prof. B. Gomard and Assistant Prof. L. Skovby, University of Copenhagen

Federal Republic of Germany W.D. Gehrmann, P.A. Maier, ERCO, Brussels, and Th. Baltes, Saarbrücken

France Prof. M. Remond-Gouilloud, Paris

Ireland M. Lynch, IIRS, and Mrs Y. Scannell, Trinity College, Dublin

Italy Prof. G. Conetti, Prof. M.L. Corbino and Prof. L. Guglielmucci, Istituto per lo Studio dei Trasporti nell'Integrazione Economica Europea, Trieste

Netherlands F. van Veen, TAUW-Infra Consult bv, Deventer

United Kingdom Prof. M. Forster, University of Southamptom

EUROPEAN RESEARCH & CONSULTING sprl

40 SQUARE AMBIORIX, Bte. 7
B-1040 BRUXELLES.

TRANSPORT OF NON-NUCLEAR, TOXIC AND DANGEROUS WASTES

TECHNICAL, SAFETY AND LEGAL ASPECTS REGARDING

PACKAGING AND MEANS OF TRANSPORT

W. D. Gehrmann

H. Dorias

H. M. C. Kaspers

P. A. Maier

November 1985

C O N T E N T S

131

LIST OF ABBREVIATIONS

ADN	Accord européen relatif au transport des marchandises dangereuses par voie de navigation intérieure = European agreement for the international carriage of dangerous goods by inland waterways
ADNR	Accord européen relatif au transport des marchandises par voie de navigation intérieure Rhin = European agreement for the international carriage of dangerous goods by inland waterways river Rhine
ADR	Accord européen relatif au transport international des marchandises dangereuses par route = European agreement concerning the international carriage of dangerous goods by road
AMvB	General Administrative Order (in Nl)
APV	General Police Regulations (in Nl)
ARV	CIM-analogue regulations for national transport by railways (in Nl)
RAM	Bundesanstalt für Materialprüfung (Federal Institute for Materials testing in FRG)
BAdS	Announcement to the Shipping (in Nl)
BGBl	Bundesgesetzblatt (German Federal Official Journal)
BR-list	British Rail list
CDG	Carriage of dangerous Goods-Committee
CIM	Convention international concernant le transport des marchandises par chemin de fer
	Since 1984 this denotation has been changed into :
COTIF	Convention relative aux transports internationaux ferroviaires

CITMD	Commission Interministérielle du Transport des Matières Dangereuses (in F)
CPL	Classification, Packaging and Labelling Regulations 1984 (UK)
CPP	Conteneurs-citernes en matière plastique non renforcée (appelation RTMD)
CSC	International Convention for Safe Containers
DB	German Railways (Deutsche Bundesbahn)
ECOSOC	Economic and Social Council (of UN)
EmS	Emergency procedures for ships carrying dangerous goods (Amendment 19-82 to the IMDG-Code)
GGVBinSch	Gefahr-Gut-Verordnung-Binnen-Schiffahrt (Inland Waterways Transportation Regulation in FRG)
GGVE	Gefahr-Gut-Verordnung-Eisenbahn (Railway Transportation Regulation in FRG)
GGVS	Gefahr-Gut-Verordnung-Strasse (Road Transportation Regulation in FRG)
GGVSee	Gefahr-Gut-Verordnung-Seefahrt (Maritime Transportation Regulation in FRG)
HGS	Handboek Gevaarlijke Stoffen (Dutch Handbook of Dangerous Substances; translation of IMDG-code with additional national prescriptions)
HW	Hinderwet (Nuisance Act in Nl)
IAEA	International Atomic Energy Agency
IBCs	Intermediate Bulk Containers
ILO	International Labour Organisation

IMDG-Code	International Maritime Dangerous Goods-Code
IM Bulk	International Maritime Bulk Code
IM Bulk Chem.	International Maritime Bulk Chemicals Code
IM Gas	International Maritime Gas Carrier Code
IMO	International Maritime Organisation
ISO	International Organisation of Standardization
JCML	Jales et conteneurs métalliques légers (appelation RTMD in F)
KCGS	Korps Controleurs Gevaarlijke Stoffen (Governmental "Corps Controllers for Dangerous Substances" of the Dutch Minsitry of Traffic and Public Waters)
LD_{50}	Lethal Dose $_{50}$
LC_{50}	Lethal Concentration $_{50}$
LDG	List of Dangerous Goods of British Rail
marg.	Marginal number in the ADR/RID-Regulations
MAK	Maximale Arbeitsplatz-Konzentration (Maximum allowable concentration in the work place)
MARPOL	Convention on the Prevention of Marine Pollution by Dumping of Wastes and other matters
MFAG	Medical First Aid Guide for use in accidents involving Dangerous Goods (Amendment 20-82 to the IMDG-Code)
MSC(DG)R	Merchant Shipping (Dangerous Goods) Regulations
N.O.S.	Not Otherwise Specified-positions
OECD	Organisation for Economic Cooperation and Development
OMI	viz IMO
PTB	Physikalisch-Technische Bundesanstalt (German Federal Physical Technical Institute)

PTT	Post, telegraph and telephone service in Nl
RGS	Reglement Gevaarlijke Stoffen (Regulations Dangerous Substances; Dutch general administrative regulation based on Dangerous Goods Act)
RICO	Règlement international concernant le transport des conteneurs (annex V CIM)
RID	Règlement International concernant le transport des marchandises Dangereuses par chemin de fer = (International convention concerning the carriage of dangerous goods by rail)
RMF	Règlement Français pour le transport par mer des marchandises dangereuses
ROSK	Reglement Ontploffingsgevaarlijke Stoffen Krijgsmacht (Regulations on Explosives of the Military Force; general Dutch administrative order on Dangerous Substances Act)
ROV	Reglement betreffende het onderzoek van vaartuigen (Dutch regulations concerning the examination of vessels); VBG annex I
RPM	Règlement pour le transport et la manutention des matières dangereuses dans les ports maritimes français
RPR	Reglement van Politie voor de Rijnvaart (the Dutch Rhine Navigation Police Regulations 1970)
RTMD	Règlement français pour le transport par chemin de fer, par voie de terre et par voies de navigation intérieure des matières dangereuses

Schepenbesluit	Ships Decree; based on IMDG-Code (in Nl)
Schepenwet	Ships-Act; based on IMDG-Code (in Nl)
Spoorwegnet	Railway Act (in Nl)
Stb or Stcr	Staatsblad/Staatscourant: official publication paper of the Dutch Government
TLV	Threshold Limit Value
UNEP	United Nations Environment Programme
VBG	Voorschriften voor het vervoer over de binnen-wateren van gevaarlijke stoffen (Dutch translation of ADNR with additional prescriptions)
VLG	Voorschriften voor het vervoer over land van gevaarlijke stoffen (Dutch translation of ADR with additional prescriptions)
VSG	Voorschriften voor het vervoer over de spoorweg van gevaarlijke stoffen (Dutch translation of RID with additional prescriptions)
WCA	Wet Chemische Afvalstoffen (Chemical Waste Act in Nl)
WGS	Wet Gevaarlijke Stoffen (Dangerous Substances Act in Nl)
WHO	World Health Organisation
WMS	Wet Milieugevaarlijke Stoffen (Bill-Act Environmentally Dangerous Substances in Nl)
WOW	Wet Overeenkomst Wegvervoer (Act concerning Agreements for Transport by Road in Nl)
WVCS	Wet Vervoer Gevaarlijke Stoffen (Bill-Act Transport of Dangerous Substances in Nl)

INTRODUCTION

Since the 70's, both the EEC and its Member States have been dealing individually with legislation on toxic and dangerous wastes. This synopsis
shall analyse the safe transportation of these hazardous materials. In the
past and even today, based on the legislation in the individual countries,
transport is handled on a sporadic basis and not always with the care
necessary to safeguard workers and the environment. The EEC has issued
some fundamental Directives; the acceleration of their enforcement and
compliance would increase the safety standards within the EEC area.

The classification of chemical wastes is very important. Originally and in
some States at the present time, the classification has been based on the
EEC-Directives 67/548; 70/189; 71/144; 73/146. This legislation does not
conform completely with the international agreements for the transport of
dangerous goods based on the UN-Recommendations on the "Transport of Dangerous Goods"; there are differences for example in classification, definition, labelling and placarding. These Directives have no influence on
packaging, because this has always depended on the specific conditions
of the traffic.

As a result, operators have often used types of packaging which were not
up to standard. This situation was the basis for research and technical
development of uniform types of packaging which would guarantee extra
safety not only in domestic traffic but also in transfrontier and transit
shipments. Naturally, this applies to normal transport situations and not
to accidents resulting from outside influences.

The above situation led to the development of special UN-Recommendations
in recent years - recommendations that form the basis for harmonizing
packaging and transportation requirements of international agreements
(RID, ADR, IMDG-Code, ADN, ICAO-Regulations). Some ratificants have
incorporated these agreements into their domestic legislation (e.g. FRG,

Netherlands, France, Italy); some countries will need some years to meet their specific legislation requirements and others have not even begun. A higher level of safety in transportation was achieved after the UN-Recommendations which developed the type and the quality of the packaging to conform with the specific dangerous properties of the transported substances (3 "groups of danger" or "degrees of danger" of substances, respectively apply to the 3 "groups of packaging"). These recommendations have been added or are in the process of being added to the relevant international conventions (ADR, RID, etc.).

The necessary uniformity pertaining to classification (criteria and properties) and the correct categorization of the relevant packaging (3 "degrees of danger"), also applies to dangerous wastes. It was achieved through the expanded and harmonized ADR/RID editions of May 1, 1985. Currently discussions are being held regarding conformity to the IMDG-Code.

For all intents and purposes, the development and the drafting of specifications and requirements has been completed with the UN-Recommendations and relevant international transport agreements. The next edition (1986/1987) of the UN-Recommendations will include a new chapter 16 (Intermediate Bulk Containers = IBCs) which will then be incorporated into specific transport regulations. In several countries there is a particular requirement of these containers (e.g. waste oil collection and transport). Proposals regarding design, construction, testing etc. for the vacuum-pressure-tanks are being drafted. The requirements may become effective in 1987.

The present synopsis has been developed on the basis of studies carried out by the relevant countries. A detailed analysis of the present situation in the EEC Member States reveals a varying degree of integration of the relevant transport requirements into national transport regulations. A few of the studies cite a transitional period of 5 years; some are very reserved regarding the time frame. Studying the relevant chapters or positions in this synopsis, it will become evident that Italy, France, the Netherlands and the FRG are on the way to a certain level of conformity.

I. CLASSIFICATION OF DANGEROUS WASTES (SUBSTANCES), TECHNICAL AND IDENTIFICATION ASPECTS

1. Definition of (dangerous) wastes

In almost all EC-member countries the definitions and interpretations of "dangerous wastes", and more specifically of "dangerous" and "waste" vary in relation to environmental laws.

In most countries waste is legally defined.
Only few countries have taken over EC definitions literally, while in other countries the existing definitions vary.
In the definitions, subjective criteria are generally combined with objective ones.
A subjective criterion is the fact that, for example, a residue has been abandoned or that a holder plans to abandon it.
An objective criterion can be the commercial value; when a residue is sold in the framework of a valorization chain it is no longer considered as waste according to the envisaged national (environmental) laws.
It is noticed, however, that the definition of the concept "waste" is subject to varying interpretations, the "hard to sell" aspect of the product being often used, in fact, more than an "unsaleable" classification.

Regarding the meaning of "dangerous" it is remarked, that not only definitions but also the applied terminology in EC-member countries differ, as well as in the different EC-Directives.
In the EC-Directive (84/631) on transfrontier shipments the term "hazardous" is used instead of "dangerous" as in other EC-Directives on waste and in international conventions and UN-recommendations. In member countries the following terms are used: chemical, special, hazardous and toxic and dangerous. With respect to the definition of "dangerous" a distinction should be made between transport legislation and environmental legislation. Whereas definitions in transport regulations only consider immediate exposure to dangerous substances, the

environmental approach takes into account the after-life of products in the environment and the harmful effects that can derive from long term exposure. In that respect the meaning of "dangerous" in transport legislation does not harmonize with the meaning of "dangerous" in environmental legislation.

In transport legislation the dangerous properties are well defined and in general based on and/or in harmonization with international regulations (UN-Recommendations, ADR/RID, etc.).

In environmental legislation, definitions of "dangerous" differ in each country.
Most are of the qualitative type, while some are quantitative, e.g. with tests for danger and with concentration levels for chemical substances. Waste lists are also applied.
It is obvious that the boundary between toxic and dangerous, and "normal" waste differs from country to country and will lead to different interpretations, by industry and authorities.

Moreover these differences in definition and interpretation may lead to all kinds of problems and misunderstandings e.g. in the trans-frontier shipments of potentially toxic and dangerous wastes, especially in relation to "waste tourism" and consequent impacts on health and the environment.

The following review sets out some of the definitions of "waste" and "hazardous waste" which are in use in various international regulations and by some organizations as well as in the respective EEC-member countries.

Transport regulations

A general outline on international transport conventions is given in annex 1 of this report.
The international transport regulations do not distinguish by classification or packaging as to whether a solution or mixture of substances is a toxic or dangerous waste or a pure chemical.

This implies that with regard to the assessment of danger it is <u>absolutely irrelevant</u> whether goods, substances, solutions, mixtures or preparations are <u>considered as being a "waste" or not</u>. In order to harmonize with the UN-recommendations of the Committee of Experts on the Transport of Dangerous Goods, the Joint Meeting on RID/ADR Regulations, March 1985, decided to add an additional sub-paragraph in the marginals 3 (4) RID/2000 (4) ADR which defines dangerous wastes, as follows: "Wastes are substances, solutions, mixtures or articles which cannot be used in their existing form but which are transported for reprocessing, recycling or elimination by burning or dumping or disposed of in some other manner".

<u>OECD</u>

By Decision and Recommendation of the Council of February 1, 1984, following definitions are given:

- "waste" means any material considered as waste or legally defined as waste in the country where it is situated or through or to which it is conveyed;

- "hazardous waste" means any waste other than radioactive waste considered as hazardous or legally defined as hazardous in the country where it is situated or through or to which it is conveyed, because of the potential risk to man or the environment likely to result from an accident or from improper transport or disposal.

<u>EEC</u>

In the Directive 78/319/EEC of March 20, 1978 definitions are

- "waste" is any substance or object which the holder wants to dispose of or is required to dispose of pursuant to the provisions of national law in force. This definition is adopted from Directive 75/442/EEC of July 15, 1975.

- "toxic and dangerous waste" means any substance containing or contaminated by the substances or materials listed in the Annex to the Directive, of such a nature, in such quantities or in such concentrations as to constitute a risk to health or the environment.

(The Annex lists 27 substances and materials constituting dangerous waste).

In the Directive 84/631/EEC of December 6, 1984, dangerous waste is defined as (art. 2);

- toxic and dangerous waste in conformity with the definition of Directive 78/319/EEC, but with the exception of the chlorinated and organic solvents as meant under point 13 and 14 of the annex to the latter Directive.

- PCB's as described in art. 1 A of Directive 76/403/EEC.

Belgium

Although in Belgium there exists no general environmental legislation on wastes (except for toxic wastes) that applies to the whole country, one national decree (Royal Decree of July, 9., 1984) has been established regarding the import, export and transit of wastes. According to this decree, wastes are defined as: "any substance or object that the owner wants to dispose of or has to dispose of on legal grounds".

Toxic wastes have been defined in the environmental law on Toxic Wastes of July 22, 1974 as: "any unused or unusable products and by-products, residues and waste generated by an industrial, commercial, craft, agricultural or scientific activity and potentially toxic to living organisms and the natural environment". The Royal Decree of February 9, 1976 further defines toxic wastes by listing different types of wastes with reference to, among other things, the toxic substances contained, the quantity and concentration of such substances and the origin of the waste.

Dangerous wastes have only been defined explicitly in the Regional Decree on Waste Disposal in Flanders of July 2, 1981. According to this decree, wastes are called "dangerous" if they expose or may expose a risk to human health or the environment or if they need to be treated, neutralized or destructed in special facilities. A list of dangerous wastes is intended to be established in future. A list of "toxic wastes" is already

used, based on the national list of toxic wastes and on the annex of
Directive 78/319/EEC.
There is also a list of "industrial wastes", which includes several types
of waste that are to be classified as "dangerous" according to the given
definition.
The list of "other" wastes includes some types of wastes, being likely
to be infectious such as certain types of wastes from hospitals, medical
or veterinary laboratories.

Dangerous wastes have been defined implicitly on several occasions, e.g.
in the regulations on dangerous goods and substances related to occupational health and safety, related to transportation, or for specific cases.
For instance in the Worker Protection Code (art. 723 bis) a classification
is given in conformity with the   6th amendment to Council Directive
67/658/EEC; in contrast with the Directive, wastes are not excluded from
these regulations.

In the regulations regarding the transportation of dangerous goods and
substances either by road, rail, sea or inland waterways, the substances
or goods that are liable to the regulations are specified in lists.
The regulations, being largely based on the respective international
agreements (ADR, RID, IMDG and ADNR) and on the UN-Recommendations on the
transport of dangerous goods, also apply to wastes.

Denmark

In the environmental legislation of Denmark, chemical waste may be described
as the special waste resulting from the production in the chemical
and related industries, as well as from the application of these products
in other firms and in households. Cf. circular on chemical waste No. 212
of October 14, 1976 issued by the Danish Department of the Environment.

As it is not possible to make a precise definition of chemical waste, 50
known kinds of chemical waste have been listed in an annex to the Ministerial Order No. 121 of March 17, 1976. Cf. Ministerial Order No. 323 of
July 3, 1980. The annex is currently being revised.

The basis of the selection of the kinds of chemical waste in question is that inexpedient disposal, etc. represents a hazard to people and environment, the kinds in question being e.g. corrosive, toxic or inflammable.

In the case of chemical waste international transport regulations are applicable, viz. ADR-regulations 1978-1985,

- ADR-regulations for road transport, with national directions for the classes 3 (in preparation) 6.1 and 8

- RID-A (Danish edition) 1985 for rail transport, with underlined national additions

- IMDG Code for sea transport with latest amendments.

For regional sea transport the following are also applicable:

- Notification No. 376 of September 30, 1983 issued by the Danish Government's Ships Inspection Service for ships under 500 G.R.T., irrespective of navigation zone, plus all ships in national traffic.

- A memorandum regarding transport of dangerous goods by ro-ro-ships.

Federal Republic of Germany

The environmental legislation in the Federal Republic of Germany regarding toxic and dangerous wastes deals with the collection, storage, disposal, treatment, carriage and tipping of these materials. This legislation is based on the Federal Waste Disposal Act of June 7, 1972 and the implementary Waste Classification Act of May 24, 1977.
Under these acts waste is defined as objects of which the owner wishes to rid himself or whose orderly removal is necessary for safeguarding the welfare of the community.

Special waste means all waste generated by industrial or commercial enterprises, which owing to their nature, composition or quantity, constitutes a particular danger to health or to the quality of air or water or which is particularly explosive or inflammable, or which contains or may produce pathogens of transmissible diseases; this waste is specifically defined in Orders.

In connection with this a list of 38 groups of hazardous and/or special wastes was issued in 1977. These wastes are identified by a five-digit system, the last two of which refer to the specific substances concerned and the industries from which they are generated.

The FRG distinguishes between transport and other possibilities of treatment. The legislation deals with treatment problems raised above and the matters of transport under the UN-Recommendations "Transport of Dangerous Goods" and other international regulations (e.g. RID, ADR, IMDG-Code) and the respective FRG legal acts (e.g. GGVSee, GGVE, GGVS).

France

In France the official concept of waste is laid down by the Act No. 75-633 of July 15, 1975 in the following manner:

"Waste consists of any residue from a process of production, processing or utilisation, any substance, material, product or more generally any movable object which is discarded, or which its holder intends to abandon."

Dangerous waste is defined mainly (on the basis of listings) by two decrees, one (of July 5, 1983) concerning "the import of toxic and dangerous waste", the other (of January 4, 1985) relating to "the control of channels for eliminating noxious waste", and coming into force on January 1, 1986.

There is also a third decree, at the planning stage, which will deal with transport. This third decree calls for the setting-up of a regional register to which companies transporting dangerous waste would have to belong. The waste products concerned would be the same as those requiring an obligatory progress schedule.

The definitions of waste in the above mentioned decrees in force are based on three types of criteria:

1. Nature of waste : the two decrees give a list of waste products considered as dangerous (this list applies the designations used in the national nomenclature of waste products);

2. Origin of waste: waste deriving from certain industries listed in the decrees is considered as dangerous;

3. Presence of dangerous substances in waste: the dangerous substances involved are listed in the decrees.

It should be noted that a strict application of these definitions may lead to certain waste products being wrongly considered as dangerous; In fact, no precise figure is given as to the dangerous substance contents that waste must contain, nor as to the nature of waste issuing from the industries listed by these decrees. Thus, waste containing only infini-tesimal quantities of dangerous substances, or unpolluted packaging waste coming from listed industries could, in theory, be considered as dangerous waste.

This lack of precision brings a need to interpret the legal texts in view of their objectives. In practice, this interpretation is carried out by the regional authorities. This being so, anomalies can appear from one region to the next as regards the classification of certain waste products.
It is obvious, that a "too liberal" approach by these subjective inter-pretations may lead to the fact that dangerous wastes are being considered as rather harmless wastes.

For international transport the international regulations are applicable.

Ireland

Toxic and dangerous wastes are defined for regulatory purposes in Ireland in accordance with the Council Directive 78/319/EEC. The wastes excluded under Article 3 of the Directive are also excluded under the Irish Regu-lations.
The exclusions extend, inter-alia, to radioactive waste, explosives, and agricultural waste of faecal origin.

National regulations on transport of dangerous substances in Ireland are generally framed in the context of dangerous goods or named chemicals rather than wastes.

Consultations have been in progress for some time between the Departments of Environment (which is responsible for the waste regulations) and Labour (which is responsible for legislation on the transport of dangerous substances by road) about the extension of the Road Transport Regulations to cover dangerous wastes, but no immediate solution is in prospect.

The basic enabling legislation (the 1972 Dangerous Substances Act) requires substances to be named individually for them to be regulated, and this may prove difficult if not impracticable in the case of certain types of complex wastes.

The ADR has not been ratified by Ireland, so the classification schemes used in that Agreement are not applicable in Ireland yet.

It is intended to expand the list of scheduled substances so that the situation in Ireland is brought closer into line with ADR requirements.

In the case of the national road transport legislation the basis of regulation is a list of specified substances. The Dangerous Substances Act, 1972 (Part IV Declaration) Order, 1980 (S.I. No. 236 of 1980) contain the list of twenty five scheduled substances which are declared by the Minister for Labour as dangerous substances for the purpose of Part IV of the Dangerous Substances Act, 1972.

The legal basis for the list is that "in the opinion of the Minister the substance constitutes a potential source of danger to person or property". The selection process for this list is understood to have included consideration of the volumes carried and frequencies of shipments of the twenty five scheduled substances by road in Ireland.

It can be seen that the list includes pure substances, solutions, and one mixture ("Any substance consisting of a mixture of butane and propane"). All the substances are either liquids or gases, no solids are included. Many of the substances are unlikely to arise as wastes, unless they are abandoned as "off-specs".

As regards rail transport the RID rules are applied, but no dangerous wastes are known to be carried by rail in Ireland. The same applies to transport by inland waterways, where there are no specific national rules on the carriage of dangerous wastes.

As regards transport by sea the relevant IMO conventions and IMDG rules are applied.

Italy

In Italy the first environmental law governing toxic and dangerous waste, enacted to implement EEC Directives in Italian legislation, was Presidential Decree No. 915 of September 10, 1982.
The definition of "waste" is set out in article 2 of that Decree.
It runs as follows:
"By "waste" is meant any substance or object derived from human activities or natural events that has been discarded or is intended to be discarded".

Waste is classified as urban, special, and toxic and dangerous. The law defines the conditions and properties determining to which of these three categories a waste is allocated.

In the case of special wastes, the quantities and concentrations that are deemed to create a danger to human health and the environment are laid down in the enclosure to the "Inter-ministerial Committee's Resolution".

"Toxic and hazardous waste" means all waste containing or contaminated by one or more of the 28 substances determined pursuant to the above Decree, and exceeding the listed quantities and concentrations of the above mentioned "Interministerial Committee's Resolution".

Regarding transport legislation, Italy follows international regulations and revises periodically its own legislation covering transport within its own frontiers according to the corresponding international regulations.

Luxembourg

According to the law of June 26, 1980 wastes are defined as: "any residue
originating from a production, transformation or utilization operation,
any substance, material, product or movable good left behind, or esteemed
so by its owner or which the owner is obliged to dispose of".

Toxic and dangerous wastes have been defined in the regulations of June
10, 1982 and include those wastes containing the substances or material
given in a list, in such conditions, concentrations or quantities as to
present a risk for causing noxious effects to the soil, the flora or the
fauna, affecting sites and landscapes, polluting the air or the waters,
generating noise and odours and in general being able to affect human health
or the environment.

A given waste is considered as toxic and dangerous if it contains listed
substances and materials in concentrations exceeding the given limits or
if it has a specific origin or may cause noxious effects.

The transportation of toxic and dangerous wastes   is   only performed by
road vehicles, it is subject to the ADR-regulations.

Netherlands

Dutch environmental legislation refrains from defining the term "chemical waste". Instead the Chemical Waste Act of February 11, 1976 calls
for three lists indicated by General Administrative Order, namely;

- a list of chemicals, divided into 4 classes of increasing concentration limits of elements and compounds
- a list of processes, which is a specific industry list
- a list of excepted substances (e.g. sewage sludge).

In the explanatory Memorandum of the Law the intention is to
cover "those materials which require special management techniques because of the potential serious damage to the health of the public or to
the environment".

It is possible that a listing-system of chemical wastes will be introduced in future. This system may have analogies with the German
and Austrian systems.

The international conventions on transport have been incorporated as a
whole in Dutch legislation. This implies   that Dutch transport
legislation harmonizes with these   conventions including classification of dangerous wastes (substances), identification of the particular
classes of risks (hazards) and criteria for handling.
Only when Dutch legislation is more stringent than the international
conventions a few additional rules comprising supplements and derogations have been made.
The differences resulting from these additional rules, however, are
marginal and most of these are dealing with transport of explosive goods
and substances.

United_Kingdom_

With regard to environmental legislation, section 30 (1) of the Control
of Pollution Act 1974 interprets. "waste" as including:

a) any substance which constitutes a scrap material or an effluent or
   other unwanted surplus substance arising from the application of
   any process

and

b) any substance or article which requires to be disposed of as being
   broken, worn out, contaminated or otherwise spoiled.

Furthermore, the Act lays down in the same section that anything which is
discarded or otherwise dealt with as if it were waste shall be presumed
to be waste unless the contrary is proved.

The Control of Pollution Act 1974, is an enabling Act permitting to make
Regulations. One such Regulation is the Control of Pollution (Special
Waste) Regulations (SI 1980/1709) which defines special waste as any
controlled waste which:

a) consists of or contains any of the substances listed in Part 1 of Schedule 1 to SI 1980/1709 and by reason of the presence of such substance

(i) is dangerous to life (in other words presenting serious hazards for a dose of 5 cm^3 ingested by a child weighing 20 kg or inhalation or contact with the skin or eyes for a period of 15 minutes)

(ii) has a flash point of 21 degrees Celcius or less

or

b) is a medical product, as defined in Section 130 of the Medicines Act 1968, which is available on prescription only

or

c) is a radioactive substance with dangerous properties other than radioactivity.

Any substance other than a medicinal product must be on the Schedule mentioned above before it may be considered as a "special waste". It is noticed that this list shows minor differences with the Annex of the Council Directive 78/319/EEC.

The potential threat to human well-being posed by these wastes is based on their inflammability, carcinogenicity, corrosivity or toxicity.

Although the Control of Pollution Act 1974 is the main and most important environmental Act covering waste disposal, there are 2 sets of regulations in force under the Health and Safety at Work etc. Act 1974 which also define hazardous wastes. In the Classification, Packaging and Labelling of Dangerous Substances Regulations 1984 (CPL) there is an "approved list" of substances regarded as dangerous for supply and for conveyance by road. The other "approved list" is issued for use with Dangerous Substances (Conveyance by Road in Road Tankers and Tank Containers)

Regulations 1981 and, so far as wastes are concerned, is identical to
the above mentioned list except for the emission (for obvious reasons)
of substance number 7013 - namely hazardous waste n.o.s. miscellaneous,
packaged.

The classification systems in use in the UK related to the transport of
dangerous wastes and substances are based on the UN classifications.
Sea-going transport follows the IMDG Code and rail transport holds its
list of Dangerous Goods (LDG). For road transport, the CPL Regulations
lay down requirements which are now in force so far as classification
is concerned.

Although the CPL Regulations, and the Special Waste Regulations referred
to above, apply to rail as well as road transport, British rail in its
LDG classifies all dangerous goods in the categories recommended by the
Committee of Experts on the Transport of Dangerous Goods.

Differences regarding classification in relation to international
transport regulations are dealt with in the next paragraph.

2. Identification of dangerous waste, criteria, hazard analysis, classification

2.1. General Remarks

With respect to the dangerous waste the rules in legislation and management techniques in the various countries are focused on characteristics and dangerous properties, treatment and disposal techniques and aspects of risk.

It will be obvious that, in general and objective terms, substances being dangerous in one country are also dangerous in another country.

Nevertheless the review of definitions in par. 1 of this chapter already shows there is no uniformity in approach and classification of dangerous waste in relation with environmental protection.

Different approaches among the various countries are marked by

- use of different characteristics
- differences in consideration and assessment of dangerous properties
- differences in risk assessment.

In many cases these approaches remain relatively vague and refer to a list of dangerous waste. Some of the main references or identification criteria for such lists are:

- the type of hazard involved (flammability, corrosivity, toxicity, irritation, etc.)
- the generic category of the goods involved (pesticides, solvents, etc.)
- technological sources (galvano-industry, oil refining, dying industry, etc.)
- presence of specific substances (PCB, dioxin, arsenic compounds).

In some countries concentration levels are set for dangerous components in waste but they vary from country to country.

For instance, Belgium and the Netherlands use the following maximum concentration levels (mg/kg):

	Cd	Hg	As	BE	CN
B	500	100	500	250	250
NL	50	50	50	50	50

It is noticed that all Member States of the EEC considered in this report have implemented in some way the Directive 78/319/CEE in their respective national legislations. Strict enforcement, however, is in most cases still in a stage of development.

It will be clear that these differences in approach have lead to different environmental legislation among the EC-member countries. Therefore under the current state of law, hazardous waste management is far from being covered by specific environmental legislation in all member countries.

The individual national transport legislations, on the contrary, show a more consistent approach in classification of dangerous substances due to the fact that they are based on international agreements.

2.2. Classification in accordance with international transport regulations

2.2.1. Comparison of national legislation and international regulations

The criteria for classification and definition of dangerous goods, substances and wastes with respect to transport are incorporated in the respective international conventions, which are transposed in national legislations. An exception is the non-application of the ADR in Ireland and of the ADNR in countries not located in the Rhine basin.

As already mentioned, with respect to the identification of the meaning of degree of danger according to the transport legislation, it is absolutely irrelevant whether goods, substances, solutions or mixtures are considered as being a "waste" or not. Environmental legislation on

the subject of toxic and dangerous substances is in most countries completely independent of transport legislation.

This implies that a waste, which is considered as a chemical, special,
dangerous and/or toxic waste in conformity with national environmental
legislation, can very well be considered as non-dangerous in conformity
with transport legislation according to international conventions and
conversely.

Since under the transport legislation no distinction is made between
goods, substances or waste, there was, until very recently, no separate
classification system for the identification of the different classes of
risk for waste, in for example RID and ARD.

Starting from May 1, 1985 the UN-classification method for solutions and
mixtures (including wastes) has been adopted for RID/ADR and is expected
to come in force in the near future. Furthermore the RID/ADR classification method is to a further extent harmonized with the UN-Recommendations.
Especially rigid changes have been made in the classes 3, 6.1 and 8.
In some other Member States other criteria as those contained in the international conventions have been set in national legislation. These
derogations will be elucidated in par. 2.3 of this chapter.

2.2.2. UN-Recommendations

The classification of dangerous goods in accordance with the Recommendations of the Committee of Experts on the Transport of Dangerous Goods is
as follows:

Class 1: Explosives
Class 2: Gases, compressed, liquified or dissolved under pressure
Class 3: Inflammable liquids
Class 4: Inflammable solids; spontaneously combustible substances; and
         substances which emit flammable gases in contact with water
Class 5: Oxidizing substances
Class 6: Poisonous (toxic) and infectious substances
Class 7: Radioactive substances
Class 8: Corrosives
Class 9: Miscellaneous dangerous substances

This classification system is elucidated in annex II of this report.

2.2.3. ADR/RID-system

Classification of dangerous goods (including wastes) and identification
of the particular classes of risk in ADR- and RID-regulations is done
according to the following classification-system:

Class 1	Explosive materials and objects	closed
Class 2	Pressurized, refrigerated liquid or	
	dissolved gases under pressure	closed
Class 3	Flammable liquids	open
Class 4.1	Flammable solids	open
Class 4.2	Materials liable to auto-ignition	closed
Class 4.3	Materials liable to generate flammable	
	gases in contact with water	closed
Class 5.1	Oxidizing substances	open
Class 5.2	Organic peroxides	closed
Class 6.1	Toxic substances	open
Class 6.2	Infectious substances	closed
Class 7	Radioactive substances	closed
Class 8	Corrosive substances	open

The products liable to the regulations are registered according to the nature of the danger in eight major classes given above as
well as in "open" and "closed" classes. For the substances or goods be-
longing to the "closed" classes, only those materials that are explicitly
registered are allowed for transportation, provided that the given conditions and prescriptions are observed. The materials or substances, even
if not explicitly listed, covered by the "open" classes are liable to the
corresponding packaging and transportation conditions given; those not
covered by the "open" classes are allowed for transportation without
special conditions; for some materials the transportation may be forbidden

according to specific provisions. Solutions and mixtures of substances which have been listed in another form or concentration than that of interest, are only liable to the regulations if they expose the same risks as the materials registered.

It is noticed, that until recently the international conventions ADR and RID only dealt with "pure chemical substances" in the classification system.

As mentioned above the criteria in the ADR and RID-conventions for classification of substances have been modified now.
The modified classification is harmonized with the recommendations of the UN-Committee of Experts. Criteria of the classes 3, 6.1 and 8 have been changed.

The new system, which was unofficially already in use for some time, has been accepted by the joint-meeting of the RID/ADR committee and brought in the integral text of the ADR and RID respectively as of 1st May 1985.

Solutions and mixtures are now defined as follows:

"Solution" means any homogeneous liquid mixture of two or more chemical compounds or elements that will not undergo any segregation under normal transportation conditions.
"Mixture" means heterogeneous materials composed of more than one chemical compound or element in the same or different aggregation.

Dangerous wastes are also included in the above mentioned definitions.
A ninth class will be probably accepted in the joint RID/ADR committee for substances which are a danger to the environment, e.g. PCB's and asbestos.
The transportation philosophy for the identification of dangerous wastes in the RID/ADR-Regulations is expected to enter into force in the near future.
The new system is explained in annex III of this report.

2.2.4. Listing of the toxic or dangerous wastes, according to the EEC Directive 78/319, in the classes as set out in the UN-Recommendations, RID, ADR and the IMO-code.

As noted previously, all member states have implemented the Directive 78/319 regarding "toxic and dangerous wastes" in their respective national legislations.
In the table of annex IV of the report for instance a comparison is made between the classifications of the toxic and/or dangerous wastes according to the Directive 78/319/EEC and the classes of the UN-Recommendations, RID, ADR and IMDG.

2.2.5. Criteria and hazard analysis

The next table gives the physical, chemical and biological characteristics (under two headings: "fundamental" and "secondary"), used as criteria for assigning substances to hazard classes.

Table: Physical, chemical and biological characteristics used as
a basis (with regard to criteria) for assignment of
substances to UN classes covered by this report
(ref. annex IV/lf)

Classes	Characteristics	
	Fundamental	Secondary
3. <u>Inflammable liquids</u> (mixtures and solutions)	fp; bp	(a) (b) (c) (d) (e) (f) (g)
4.1 <u>Inflammable solids</u> (mixtures)	fp; (d)	(c) (s)
5.1 <u>Oxidizing substances</u> (mixtures and solutions)*	(m) (g)	(a) fp (b) (c) (d) (h) (f) (i) (l)
5.2 <u>Organic peroxides</u> (mixtures and solutions)	(r)	(a) fp (b) (c) (d) (m) (e) (h) (f) (g) (n) (i) (o) (l)
6.1 <u>Toxic substances</u> (mixtures and solutions)	bp (i)	(a) fp (b) (c) (d) (m) (h) (p) (f) (o) (p) (l)
8. <u>Corrosive substances</u>	(h)	(a) fp (b) (c) (d) (p) (m) (f) (i) (o) (q) (l)
9. <u>Dangers not covered by other classes</u> (mixtures and solutions)	(f) (a)	(b) (h) (g) (l) (i) (o) (q)

* Mixtures with explosive substances may not be transported if
they might explode in contact with flame.

Legend:

fp	flash point
bp	boiling point
(a)	vapour pressure
(b)	inflammability
(c)	explosive properties
(d)	self-inflammability
(e)	viscosity
(f)	mixibility with water
(g)	unstable substance content
(h)	corrosivity
(i)	acute toxicity
(l)	irritation (skin, eyes, mucous membrane)
(m)	oxidizing properties
(n)	cubic expansion on temperature change
(o)	chronic toxicity
(p)	chemical reactions with water
(q)	genetic effects
(r)	self-accelerating decomposition temperature
(s)	melting point

The use of these characteristics for assignment is elucidated in annex V of this report.

2.3. Derogations and supplements in national transport legislation with regard to international transport regulations for classification of dangerous waste.

2.3.1. Integration of international regulations in national transport legislation

International transport regulations have been integrated in national legislation in most cases.

National regulations comparable to these international regulations are listed in the following table.

Table

Transport by	Road	Railway	Inland waterway	High sea and sea harbours
International regulations	ADR	RID	ADNR	IMDG
Belgium	(2)	(2)	(2)	(2)
Denmark	(2)	(2)	(2)	(2)
FRG	GGVS	GGVE	GGVBin Sch	GGVSee
France	RTMD	RTMD	RTMD	RMF
Ireland	(2)	(1)	(1)	(2)
Italy	(8)(9)	(2)(7)	(12)(13)	(10)(11)
Luxembourg	(2)(6)	(1)	(1)	(1)
Netherlands	VLG	VSG(13)	VBG	HGS
United Kingdom	CPL	LDG(13)	(3)(4)(13)	(4)(5) IMDG

Footnotes

(1) No relevant information or no transport by this means

(2) References are made to the international regulations

(3) Terms and Conditions for the Transport of Dangerous Goods on the Board's Waterways and Docks

(4) Dangerous Substances in Harbours and Harbour Area Regulations of 1985

(5) Merchant Shipping (Dangerous Goods) Regulations

(6) Transport of toxic and dangerous wastes being only performed
     by road vehicles

(7) Enclosure 7 of "Conditioni e Tariffe per i transporti delle
     cose sulle Ferrovie dello Stato"

(8) Presidential Decree 895 of November 20, 1979

(9) Presidental Decree 343 of June 15, 1959

(10) Presidential Decree 1008 of May 9, 1968

(11) Ministerial Decree of May 15, 1972; Gazetta Officiale 214 of
     August 18, 1972

(12) ADN, Committee Resolution 233 of February 6, 1976

(13) Only very few transports of waste by this means

2.3.2. Evaluation by country

Belgium

Most international conventions, including ADR/RID, ADNR and the IMDG-
Code, have been adopted in Belgian law.

For ammonium nitrate and certain mixtures of it, specific regulations
have been established regarding the transportation, the storage and the
sale in view of its explosive potential. Various types of mixtures being
liable to the regulations have been defined.

Other exceptions to the international regulations being relevant to
wastes exist in the class 3 - inflammable liquids - of the ADR. The
exceptions concern liquids of the divisions 2° to 5°, except carbon
disulphide, ethyl ether, petroleum ether, pentane and methyl formiate as
well as acetaldehyde, acetone and acetone mixtures being packed in small
receptacles. Another exception concerns extra fueloil, category E -NBN
52-501- on the condition that its temperature is 8°C lower than the
flashpoint of the material.

France

With regard to transportation by road or rail, international transport
is not subject to RTMD (like national transport), but is governed by
the RID and ADR conventions (art. 8, RTMD). The French and international
nomenclatures are in any case very similar, especially in the matter of
categories 3, 6.1 and 8 (which cover the great majority of substances and
waste transported), since the latest revision of the RTMD (due to come
into effect on January 1, 1986). In fact, the classification criteria
for substances in these categories and the corresponding danger groups
are now very similar in comparison with international regulations and the
RTMD.
In general, the UN numbering of substances is repeated by the RTMD.
Where this is not the case, the identification number of the substance
in question is preceded by the figure 9, indicating clearly that this
item is specific to the French regulations.

Navigation on the Rhine is governed by the regulations for the transport
of dangerous goods on the Rhine (ADNR), made applicable in France by a
decree of December 8, 1971.

As regards sea transport, the only differences which may appear in the
classifications are due to recent amendments to the IMDG-Code which have
not yet been taken into account by the RMF.
It may also be noted, that foreign ships operating in French ports need
not to conform to RMF standards if they conform to the IMDG-Code.

Ireland

The ADR Agreement has not been ratified by Ireland, so the classification scheme used in that Agreement is not applied in Ireland. The
prospects for extension of the list of dangerous substances are very
good, but petroleum and explosive substances may still be covered by
separate legislation. Moreover, the national enabling legislation may need
to be changed to allow transport of classes of dangerous substances to be
regulated even where the exact chemical composition is complex and
difficult to determine (as in the case of some dangerous wastes).

In the opinion of the national reporter, the national regulatory system relating to the transport of dangerous wastes is inadequate. The basic problem is that the transport of these wastes is not adequately controlled by the national regulations on dangerous wastes, and the national Road Transport regulations do not apply in practice to dangerous wastes. The answer to the problem clearly lies in extending the list of scheduled substances under the Road Transport regulations in order to include dangerous wastes. This may require an amendment of the original 1972 Act in the way that substances can be regulated on the basis of their hazard characteristics without the substances having to be pure and even without their having to be fully chemically characterised. It is understood that work is in progress towards such a broadening of the legislation.

The current dichotomy between the Road Transport regulations and Dangerous Waste regulations leaves a serious gap in the national system of control of dangerous wastes.
There are indications that the Government wants to move towards ratification of the ADR Agreement, but this will require a considerable expansion of the scope of current regulations, and will impose considerable costs on both industry and the State if compliance with the ADR is to be effectively monitored and enforced. The cost will arise particularly in meeting and policing ADR specifications.

The national system of regulation in regard to rail transport of hazardous wastes suffers from similar deficiencies to the Road Transport regulation system. The basic problem again is the very restricted list of substances covered by rail regulations.
Moreover, the regulations in the case of rail transport are still only in draft form.
Rail transport is not used for dangerous wastes in Ireland, so the problem of regulatory inadequacy is more notional than real. The same is true of inland waterway transport, where there are no known transport operations involving dangerous substances.

Italy

For international transports by rail the RID-regulations are followed.
For inland transports, however, the regulations of
"the Terms and Tariffs for the carriage of goods on State Railways" are
followed.
Within this context a classification list of 15 categories of goods
and six technical appendices is used. This classification system is
not completely alligned with the RID-classification (8 categories).
For international transport the ADR regulations are followed.

For inland transport the law of August 10, 1970 refers to classification
but leaves the possibility of complying with ADR. Road transports have
their own regulation (Highway Code), different from the ADR.

For transport by sea regulations are issued by the Italian Ministry of
Merchant Marine. These regulations are in line, as far as possible, with
the IMDG-Code. Only very little transport is realized by inland waterways.

The Netherlands

Dutch legislation regarding transport of dangerous wastes contains no
relevant differences from international conventions. With respect to
classification, only small derogations occur (for instance a slightly
different text in some margin numbers of ADR/RID-regulations).
In principle, the classification of goods, substances or wastes is
carried out by the person who offers the goods, substances or wastes
for transportation.

In case classification is difficult, competent authorities may be consulted. In very difficult cases a classification problem can be brought
before the "Classification Committee of Experts" (Rubriceringscommissie)
of the Ministry of Transport and Public Work. Classifications made
by this committee are compulsory. The committee reviews mainly unknown
or new goods and substances for which classification cannot be simply
carried out on the basis of the criteria of ADR, RID, ADNR and IMO.

By Ministerial Decree this committee may allow exemptions from the conditions, which result from the classification procedures, and may also allow incidental transports which normally would have been forbidden under the rules of the transport legislation.

These exemptions are made possible for well defined circumstances, e.g. when the prescribed conditions (packaging, labelling, etc.) cannot be matched in reasonable terms.
This can be the case when for instance certain wastes have to be transported and the costs of the prescribed (new) packaging are far too high to be economically feasible. Prescriptions for other safe packaging methods which can differ from international conventions may then be given.

In case analysis of samples is required, this can be done by the person himself or by a contracted laboratory.

United_Kingdom_

- General

The classification systems in use in the UK related to the transport of dangerous wastes and substances are based on the UN-classification.
Sea-going transport follows the IMDG-Code.
Rail transport has a List of Dangerous Goods (LDG).
For road transport, the CPL Regulations lay down requirements which are now in force so far as classification is concerned.

Hazardous wastes carried in deep sea vessels are subject to the International Maritime Dangerous Goods Code, as amended from time to time, made under the chapter VII of the 1974 convention for the Safety Of Life At Sea (SOLAS), which came into force in the UK on May 20, 1980 although some conditions imposed by the Merchant Shipping (Dangerous Goods) Regulations 1981 relating to packaging will remain in force over the transitional period ending on January 1, 1990.

Special waste status is accorded to a waste under the Control of Pollution (Special Waste) Regulations 1980, if a single dose of not more than 5 cubic centimetres of the waste would be likely to cause death or serious damage to tissue if ingested by a child of 20 kg body weight (the approximative weight of a five-year-old). In addition, waste is to be regarded as dangerous to life if exposure to it for fifteen minutes or less would be likely to cause serious damage to human tissue by inhalation. From time to time the Department of Environment publishes waste management papers giving guidance to the various regulations.

- Road, rail

The CPL Regulations 1984 do not use the class numbers used in ADR. All eight ADR classes are included in the CPL Regulations, but the regulations have a ninth class, "other dangerous substances", which may include asbestos, etc., known to be harmful but which do not yet fit in any of the existing ADR classes. (Note: a ninth ADR/RID class, which also includes asbestos and PCB, may be introduced). There are differences in Class 6.1 (ADR) relating to toxicity, where, although both systems use three categories of toxicity, that for ADR being less strict than the UK classification, except for toxicity and inhalation, where both systems are the same for highly toxic (very toxic - UK) and toxic substances.

Similarly, differences exist in Class 3 (flammable liquids), where in the UK an extremely flammable substance must have a flashpoint below 0 degrees Celsius.

Flammable substances in the UK have a flashpoint below 55°C, but are subject to ADR with a flashpoint up to 100°C.

In diagram:

Classification of flammable liquids (degree Celsius)

Regulation	Extremely flammable	Highly flammable	Flammable
CPL (supply)	<0	0 - 21	21 - 55
CPL (conveyance)			<55
BR-List		<21	21 - 55
IMDG	<18	-18 - +23	23 - 61
UN		<23	23 - 60,5
ADR/RID		<21	21 - 100

The CPL Regulations consider a non-corrosive substance dangerous for supply if, through immediate, prolonged or repeated contact with the skin or mucous membrane, it can cause inflammation.

Other legislation is silent on these aspects, but clearly almost all corrosive substances (see below) have the potential in their dilute forms to cause skin irritation.

According to the IMDG-Code corrosive substances are solids or liquids possessing (in their original state) the common property of being able, to a greater or lesser extent, to damage living tissue. The CPL Regulations for supply regard any substance as corrosive which may, on contact with living tissues, destroy them. That part of the regulations which deals with conveyance is a little more specific, describing a corrosive substance as one which, by chemical action, will cause severe damage when in contact with living tissue. The Special Waste Regulations regard waste as dangerous to life if exposure to it by skin or eye contact for fifteen minutes or less would be likely to cause serious damage to human tissue. Serious damage is considered to be permanent disfigurement or physical impairment and may be localized or, on absorption into the body, have a systemic effect. It is important to observe that serious damage to the eyes or oesophageal tissues is likely to occur at much lower concentrations than those which are required to cause skin damage generally.

Although the CPL Regulations, and the Special Waste Regulations, apply to rail as well as road transport, British Rail in its LDG classifies all dangerous goods in the categories recommended by the Committee of Experts on the Transport of Dangerous Goods.

Each category, at the moment, is subject to the provision that wastes containing substances of particular classes of risk must be accompanied by the correct consignment note (until the repeal of Deposit of Poisonous Waste Act 1972). The LDG is undergoing continual revision and one of the amendments proposed for the near future is to bring it up to date by substituting the requirements under the Control of Pollution (Special Waste) Regulations 1980 for those of the 1972 Act.

Another implementation of the UN-Recommendations is the provision in each list of the required packaging for rail transport.

There is also a working manual for British Rail Staff, of which a part (the "pink pages") relates to the handling and conveyance of dangerous goods. The pages specifically include wastes similar to those described in the approved list under the CPL Regulations.

The "pink pages" cover classification of dangerous goods, marking, loading and unloading, marshalling and movement as well as working considerations such as reporting procedures for incidents involving dangerous goods.

Comparing British Rail's LDG with RID, the main difference is that the BR list has two extra classes not present in RID, that is Class 9 (miscellaneous dangerous substances) and Class 10 (dangerous chemicals in limited quantities). Class 9 includes asbestos and Class 10 is for substances of more than one classification carried in the same package.

In Class 6.1, toxic substances, three degrees of toxicity are specified in RID, with LD_{50} and LC_{50} values identical to those in ADR, except for harmful solids with oral LD_{50} values between 50 and 200 mg/kg, both of which are subject to RID but not ADR. In British Rail's LDG there are only two categories, with listed substances, although LD_{50} or LC_{50} values are quoted.

The LMDG-Code (Class 6.1) describes as toxic (or poisonous) substances
liable either to cause death or serious injury, or to harm human health
if swallowed or inhaled, or by skin contact.

The LDG is rather more detailed, in that it classifies as toxic (Class 6a)
all those liquid or solid substances which, in the opinion of the British
Rail Board, are sufficiently toxic or require labelling to indicate the
nature of the potential hazard during transport. Substances which present
some other hazard, in addition to toxicity, may be placed in the alternative class. Because of the wide range of toxic hazards normally displayed by poisonous weedkillers, pesticides and insecticides will not be
carried by BR until they have been classified.

3. Test-facilities

Various reports indicate addresses concerning testings and/or test-facilities. No test-facilities are known in Luxembourg. There may be limitations in laboratory capacity and scope in Ireland.

II. PACKAGING

1. Types and conditions

1.1. General packaging requirements

Until the sixties, it was usual that the packaging of every commodity was
individually constructed and built; this was especially important for dangerous goods. On the other hand, this method of manufacture became more
and more expensive due to the variety of products. In addition to this
was the danger of unexpected accidents caused by inadequate or inappropriate packaging of the dangerous goods.

Based on this experience, the "Group of Rapporteurs" and the "Committe of
Experts" of the "Economic and Social Council (ECOSOC)" of the UN developed
a new packaging philosophy in the sixties and seventies.

The basic principle was to standardize packaging which could be used for
a great variety of dangerous goods. The basic requirement was that these
goods have similar properties, criteria and degrees of danger. There must
be a conformity between the relevant "packaging group" and the "group*
of danger" of the substance. This fundamental philosophy makes standardized packaging usable for the common hazardous materials as well as
for toxic and dangerous wastes. This simplifies the regulations and the
procedures, because is was necessary to create special regulations for
wastes.

The result is outlined in chapter 9 (General Recommendations on Packaging)
of the UN-Recommendations "Transport of Dangerous Goods" and the harmonized
RID/ADR-Regulations concerning Annex V of RID and Annex A 5 of ADR. Similar
modifications took place with the construction, performance testing, marking and operating procedures of tanks and tank-containers.

---

* degree of danger

As mentioned above this UN philosophy was the basis for the decision of the RID/ADR-Joint Meeting (March 1985) to decide against any special treatment concerning classification and packaging of dangerous wastes. They are to be classified and packed as normal chemical products or their solutions and mixtures.

The "EEC-group of experts for the transport of dangerous goods" and the "RID-Safety-Committee" also approved these decisions without essential differences.

These RID/ADR alterations relating to international law of land-transport of dangerous goods came into force on May 1, 1985. The mentioned developments of harmonization have pendants in other organisations of transport such as IMO.

Belgium refers to the "Worker Protection Code" *) (art. 723bis). According to this Code, the receptacles, sacs and wrappings containing dangerous goods or preparations, as well as their closure systems are required to show perfect tightness and rigidness in order to guarantee the containment of their content during handling.

This Code is to be seen too in connection with the Royal Decree of February 9, 1976 (List of toxic wastes) which stipulates that the transportation and the packaging of the waste has to be performed in such a way as to exclude any danger of any pollution caused by the transportation. There is no legislation relating to RID/ADR-regulations (without test procedures).

---

*) Based on EEC-Directives on "Classification, Packaging and Labelling of dangerous goods"

Federal Republic of Germany

The developments *) in the international legislation for the transport of
dangerous goods by rail and road came into force in the FRG on July 30,
1985 (Decree on the interstate and trans-frontier transport of dangerous
goods by road (GGVS), July 22, 1985; BGBl. I, S. 1550 / Decree on the
interstate and trans-frontier transport of dangerous goods by rail (GGVE),
July 22, 1985; BGBl. I, S. 1560).

The relevant requirements of the Gefahr-Gut-Verordnung-See (GGVSee) are
nearly identical to these of the IMDG-Code and the Gefahr-Gut-Verordnung-
Binnen-Schiffahrt (GGVBinSch) to those of the ADR.

France

The requirements and codes of the RMF are taken as being identical to
those of the IMDG-Code. The land-regulations are similar to RID and ADR.
There are only a few differences, mainly concerned with tanker vehicles,
i.e. the French regulations set stricter requirements (in conformity with
FRG requirements). Moreover, the RTMD, unlike the international regulations, lays down special requirements for tanker vehicles intended solely
for the transport of wastes. Similar requirements in RID/ADR are in preparation and will probably come into force by the end of 1987.

The above mentioned recent requirements of RID/ADR are subject to incorporation into special French legislation.

Ireland

ADR has not been ratified by Ireland, so the classification scheme used
in this agreement is not applied in Ireland, and the special RID/ADR-
procedures for the transport of dangerous goods would be almost impossible
to carry out (including packaging performance tests).

The "Toxic and Dangerous Waste Regulations" of 1982 will be among other
things mandatory for packaging.

---

*) harmonization RID/ADR (1.5.85)

Italy

The different international Regulations for the international transport
of hazardous goods are in force in Italy.

Netherlands

The specific international requirements of packaging (RID/ADR etc.)
relating to all modes of transportation apply in the Netherlands with
additional regulations for some chemicals.

United Kingdom

For road transport, the only relevant legislation applies to loads exceeding 3 cubic metres in volume carried in tanks or tank-containers and is
provided in the "Dangerous Substances (Conveyance by Road in Road Tankers
and Tank-Containers) Regulations 1981" for the testing and examination of
tanks, whether fixed to road tankers or removable. The performance testing
of other packaging used for road transport is not required at present
under British law, and there does not appear to be any likelihood of such
legislation being introduced in the foreseeable future.

For rail transport, the packaging requirements are laid down in "British
Rail's List of Dangerous Goods (LDG) and Conditions of Acceptance".

UK will accept for carriage, dangerous goods which are contained in receptacles or packaging which have passed the UN or the RID/ADR tests. Similar is the situation for transport by sea.

1.2. Types of packaging (codes and specifications)

1.2.1. Packaging with a capacity not exceeding 450 l or 400 kg net
        weight *)

Toxic and dangerous wastes must be packed in receptacles of good quality
that are constructed and so closed as to prevent the package prepared for
shipment from leaking which might be caused, in normal conditions of transport, by a change of temperature, humidity etc. These provisions apply to

---

*) Chapter 9 / Un-Recommendations; Annex A 5/ADR; Annex V/RID

new receptacles which are re-used. Before the re-use, every package must be inspected and must not be used again unless free from corrosion and other damage.

Packaging (including closures) in direct contact with hazardous materials must be resistant to any chemical or other action of such goods; the material of the receptacles must not contain substances which may react dangerously with the contents, form hazardous product or significantly weaken the receptacles. Nevertheless, it is the responsibility of the shipper to ensure that such packaging is, in every way, compatible with the dangerous wastes to be contained within such packaging. This particularly applies to corrosivity, permeability, softening, premature aging and embrittlement. The closure device must be so designed that it is unlikely that it can be incorrectly or incompletely closed, and must be such that it may be checked easily to determine that it is completely closed.

Dangerous substances or wastes of all classes other than the Class 1 (Explosives), 2 (Gases), 6.2 (Infectious Substances) and 7 (Radioactive Substances) have, for packaging purposes, been divided among 3 groups according to the degree of danger they present:

Packing Group a) of RID; ADR; ADNR or 1) of UN; IMO : <u>great danger</u>
Packing Group b) of RID, ADR; ADNR or 2) of UN; IMO : <u>medium danger</u>
Packing Group c) of RID; ADR; ADNR or 3) of UN; IMO : <u>minor danger</u>

The packaging permitted for transport of dangerous goods and wastes must comply with the specifications in table IV, pass the performance tests, and be UN marked.

In the UN-Recommendations, IMDG-Code, RID/ADR-Regulations and ICAO-Regulations, codes are allocated to most packaging. The following Arabic numerals indicate the types of packaging :

1. Drum

2. Wooden Barrel (not used in these matters)

3. Jerrican

4. Box (not used in these matters)

5. Bag (not used in these matters)

6. Composite Packaging

7. Pressure receptacle (vessel)

The following Latin characters indicate the material:

A. Steel (all types and surface treatment)

B. Aluminium (not used in these matters)

C. Natural Wood (not used in these matters)

D. Plywood (not used in these matters)

F. Reconstituted Wood (not used in these matters)

G. Fibreboard (not used in these matters)

H. Plastic Material

L. Textile (not used in these matters)

M. Paper, multiwall (not used in these matters)

N. Metal (other than steel or aluminium)

P. Glass, Porcelain or Stoneware (not used in these matters)

Table IV

specifications, types and codes *)

type	material	code	category	capacity
1. Drums	A. Steel	1A1	non-removable head	450 l 400 kg
		1A2	removable head	
1. Drums	H. Plastics	1H1	non-removable head	450 l 400 kg
		1H2	removable head	
3. Jerricans	A. Steel	3A1	non-removable head	60 l 120 kg
		3A2	removable head	
3. Jerricans	H. Plastics	3H1	non-removable head	60 l 120 kg

---

*) can be used in these matters

In legislation and practice there is a conformity within the Netherlands, Italy, France and the FRG.

Ireland has not ratified the ADR and the specific requirements of packaging in the national decrees are too general. They should correspond to those of ADR, but do so for only 25 substances for road transport. The requirements in regard to dangerous wastes are also much less specific or detailed, with no conformity to the above mentioned recent agreements of the ADR/RID Joint Meeting (March 1985).

The British types of packaging have only a small degree of conformity with requirements in the RID/ADR-Agreements and those in chapter 9 of the UN-Recommendations.

Within world-wide trade there is a movement among industries towards greater conformity with international regulations related to the requirements for the design, construction and testing of hazardous materials as well as availing of the technological advancements in packaging. As a result, on January 1, 1984 new arrangements for the testing and marking of packaging were introduced in the UK. This was to ensure that all packaging, irrespective of the mode of transport proposed, was manufactured as recommended by the UN Committee of Experts on the transport of Dangerous Goods in respect to the relevant RID/ADR specifications.

The required tests will be carried out in accordance with the relevant parts of ISO Standards 2248, of which the British equivalent is BS 4826. It is expected that by 1990 all packaging will be tested according to UN standards; however, until that date, the requirements laid down by the British Standards Institution will continue to be met, in parallel with the UN specifications.

1.2.2.   Intermediate Bulk Containers (IBCs) UN-Recommendations;
         (new chapter 16)

1.2.2.1.   Flexible IBCs

Flexible IBCs are manufactured (with or without plastic lined textiles)
into bags with a capacity of 450 - 1000 litres. They are used for the
transport and storage of solids. For the carriage of toxic and dangerous wastes and goods, this type of container may be used only as an
exception. That is the opinion expressed in the studies of all Member
States.

1.2.2.2.   Metallic (prismatic) IBCs

This type of container with a capacity of up to 3000 litres is mostly of
cubic design, is stockable, economical, and safe for carriage, storage
in the chemical industry. These containers have a normal pallet design
and can be moved by forklift or crane and, with exceptions, are permitted
for use for the Classes 3; 4.1; 4.2; 4.3; 5.1; 6.1 and 8.

It is excellent in handling of dangerous wastes, e.g., collecting,
storing, and transport to treatment facilities. A typical example for
this usage is the collection of waste oil. In some Member States such
as the UK or the FRG interesting developments can be observed.

The UK does propose and use IBCs produced by rotational moulding of powdered
thermoplastic materials. The "Dyecon 1200" tank, produced especially for
hazardous goods, probably wastes too, is a 1200 l unit fitted with skids for
stacking. An ultra-violet inhibitor in the plastic mix enables the unit to
be used outdoors. From the same manufacturer the "Contitainer 800" is an
800 l plastic tank encased in a steel frame for hazardous goods. It can
also be used for wastes. There is conformity with France, FRG and the
Netherlands.

1.2.3. Containers

Containers mean re-usable units generally with a cubic shape, designed
and constructed to permit lifting with their contents intact and are primarily for containment of packages of chemical products and dangerous
wastes during transportation. They are standardized by ISO (International Organisation for Standardisation). Within the "International Convention for Safe Containers" (CSC) the State parties of the agreement are
liable, that containers in trans-frontier shipment are tested, approved
and the result certificated on a metal plate permanently attached to the
shell. Wellknown types measure 20 or 40 feet.

Other containers are common tank-cars or hopper and tipper vehicles
for road and rail. The factors to dictate the selection of the containers
include size of shipment,   compatibility of the product with the container material, ease of storage, handling and cost.

1.2.3.1. Legislation and use in the Member States

In all Member States, the "CSC" convention applies. Beyond it, Italy has
adopted the UIC requirements 590 too (UIC: Union Internationale des Che-
mins de Fer). The situation in France is similar. According to UIC 590,
their authorization is issued by the "Container Bureau of the SNCF". In sea
transport (RMF, Annex C) special requirements for containers are being
worked out.

1.2.4. Tank-containers

A tank-container is defined as a tank having a capacity of not less than
450 litres whose shell is fitted with the items of service and structural
equipment necessary for the transport of dangerous liquids, gaseous or
powdery/granular solid substances or wastes. The tank-container should be
capable of being carried by land and sea and of being loaded and discharged
without the need to remove its structural equipment. It should possess
stabilizing structures external to the shell, and should be capable of
being lifted when full. It should be transported only on those road or
rail vehicles whose fastenings are suited to conditions of maximal permissible loading of the tank-containers.

With regard to the universal usage of a multimodal tank transport it has been named the Multimodal Tank-container. In the chapter 12 of the UN-Recommendations on the "Transport of Dangerous Goods" (Recommendations on Multimodal Transport) the design, construction, testing etc. are required as basis for all relevant international agreements.

The tank-containers for sea transport in IMO are regulated as so-called IMO-types (IMO-type I, II, IV and V are used). For land transport the tank-containers are required in Annex B.1b/ADR and Annex X/RID.

In the Netherlands, the FRG, France, Italy the specific requirements in the international agreements apply in their domestic regulations.

1.2.5. Tanks

A tank is any tank which is used for the conveyance by road or rail of a liquid, gaseous, powdery or granular material or sludge or waste, in bulk, and so constructed that it can be securely closed (except for the purpose of relieving excessive pressure) during the course of the carriage. Any tank can be permanently attached to these vehicles and can be loaded or unloaded without being removed from those vehicles.

The relevant requirements in Annex B, 1a deal with construction, equipment, testing, approval, operation, etc. of the tanks. They should be made of metallic materials.

With the recent revision of ADR, the new created Annex B. 1c involves tanks made of reinforced plastic. It outlines their construction principles as well as the criteria for tests and evaluation regarding safety and technical feasibility. Therefore, according to ADR, certain toxic and dangerous substances are allowed to be transported in such tanks.

Similar to Annex B. 1a/ADR (metallic tanks) is the Annex XI/RID with the specific requirements of construction, testing and operation of tank-wagons by rail.

In Belgium, the FRG, France, Italy, the Netherlands and the UK the specific RID/ADR tank regulations apply.

1.3. Consignment procedures

Whenever dangerous goods are offered for transport, certain measures
should be taken to ensure that the potential risks are adequately commu-
nicated to all who come in contact with the goods in the course of transport. This has traditionally been accomplished through special marking and
labelling of packages to indicate the hazards of a consignment and through
the inclusion of relevant information on the documents which accompany the
shipment and by placarding of the transport units.

1.3.1. Marking of packaging

The marking indicates that the packaging which bears it corresponds to a
successfully tested design type which is related to the manufacturer, but
not to the use of the packaging. In itself, therefore, the mark does not
necessarily confirm that the packaging may be used for any substance: generally the type of packaging (e.g. steel drum), its maximum capacity and/or
mass, and any special requirements are specified for each substance in the
regulations for each mode of transport.

The marking is a guarantee that all packaging will correspond to the
tested prototypes over the whole period of use. Each packaging intended
for use should bear durable and legible markings as follows:

a) The United Nations packaging symbol:

b) The code number designating the type of packaging according to the
   sub-para. 1.2.1.

c) A code in two parts:
   (1) a letter designating the packing group(s) for which the
       design type has been successfully tested:
       X for packaging group I, II and III (degree of danger: a), b) and c))
       Y for packaging group II and III (degree of danger b) and c))
       Z for packaging group III only (degree of danger III)

(2) For packaging intended to contain liquids, this letter is followed by the relative density if it is greater than 1,2; for packaging intended to contain solids or packaging containing inner packaging, the letter is followed by the letter S.

d) The last digits of the year during which the packaging was manufactured (for plastic packaging, the month of the manufacture should also be marked on the packaging, but it may be marked in a different place from the remainder of the marking):

e) The State authorizing the allocation of the mark, indicated by the distinguishing sign for motor vehicles in international traffic.

f) The name of the manufacturer or other identification of the packaging specified by the competent authority.

Example of marking: 1A1/X 1,6/400/85
                              I / ISP-FART748

This is the marking for a steel drum (1A1) which may be used to transport liquids in Packing Groups I, II and III=(X) with the density of not over 1,6 and which has been subjected to a hydraulic pressure test with a gauge pressure of 400 kPa.

1.3.2. Labels identifying risks

The labels are diamond-shaped pictograms and mainly intended for affixing to hazardous materials in packaging, tanks, vehicles etc... The labelling system is based on the classification of dangerous substances and was established with the following aims in mind *):

------

*) It must be differenciated between labels used world-wide for the transport of dangerous materials by all modes of conveyance and the "Workers Protection Labels", proposed in the EEC-Directive 67/548 and enforced in all EEC Member States for the usage in industrial facilities.

a) to make dangerous substances or wastes easily recognizable from a
   distance by the general appearance (symbol, colour and shape) of the
   labels they bear;

b) to make the nature of the risks easy to identify by means of symbols:
   Bomb (Explosion), Flame (Fire), Skull and Crossbones (Poisons), Trefoil
   (Radioactivity);liquids spilling from two glass vessels and attacking
   a hand and a metal (Corrosives) are supplemented by non-inflammable
   compressed gases (a gas cylinder). Infectious substances which should
   be stored away from foodstuffs (St. Andrew's cross over an ear of wheat);
   and

c) to provide, by means of colours on the labels, a useful guide for handling and stowing.

It should be pointed out that:

a) for transport by sea, the type of danger must be indicated in the
   lower half of the label;

b) for transport by land, a single label is used for explosives (an explo-
   ding bomb) and there are no specific labels for Class 2.

The FRG, Ireland, Italy, Luxembourg, the UK, France and the Netherlands are
in conformity with the above mentioned specific requirements.

According to the Belgian legislation each packaging of toxic wastes has
to be provided with an indication which identifies its content. With
specific transport-requirements they would be labelled according to the
classified dangerous goods, which are in general not likely to apply to
wastes. It may be required on the basis of the "Worker Protection Code"
(art. 723bis). These principles are the same as proposed in the EEC-
Directives regarding "classification, packaging and labelling of dangerous substances and preparations".

According to Ministerial Order No. 323 of July 3, 1980 and Ministerial
Order No. 121 of March 17, 1976, of the Danish Department of the Environment, the Danish National Agency of Environmental Protection prepares
chemical waste cards on which under 7 the requisite marking is stated.

1.3.3. Documentation of dangerous wastes shipment

One of the primary objectives of a shipping document for dangerous goods
or wastes is to convey the fundamental information relative to the hazard
of the materials being transported. To achieve this, it is considered
necessary to include certain basic information on the transport document
concerning the hazardous substance. It is recognized that in case of
wastes, individual national authorities (e.g. provinces, counties, departments, Länder) or international organisations may consider it necessary
to require additional information. However, the basic items of information considered necessary for each hazardous material or waste offered
for transport, by any mode, are:

a) the proper shipping name;
b) the class or, when assigned, the division of the substance;
c) the UN serial number (also N.O.S.-number) assigned to the substance
   or article or the identified waste;
d) the total quantity of the hazardous materials or wastes covered by
   the description (by volume or weight as appropriate).

If dangerous wastes are being transported, the proper shipping name should
be preceded by the word "Waste".

1.3.3.1. Situation in the EEC Member States

For the transport by rail, road, inland waterways, on sea, the relevant
requirements of RID, ADR, ADNR, IMDG-Code apply.

The Royal Belgian Decree of September 11, 1967 stipulates that a document,
written in at least one of the national languages (Dutch, French, German),
must be filled in by the consignor in case of transportation of materials for business.

In Denmark, dangerous waste shipment is documented in section 1 and 5 and
sub-section 2 of section 8 of Ministerial Order No. 121 of March 17, 1976.

Beside the relevant transport requirements in the FRG, according to the
Länder laws and the EEC-Directives, specific waste information must be
added.

It should be noted that "the declaration of shipment of dangerous goods" prescribed by the RMF of <u>France</u> is almost identical to the "declaration of loading of dangerous goods" required by the RTMD for any transport of goods by land; indeed, these two declarations can be made on the same document.

In <u>Ireland</u>, the specific documentation of dangerous waste shipment is Article 8 of the National Regulations on Toxic and Dangerous Wastes (S.I. No. 33 of 1982) which deals with the nature, content and processing of the Consignment Notes which must accompany all shipments of toxic and dangerous wastes.

1.3.4. <u>Placarding</u> (i.e. orange placard)

Placards should be affixed to the exterior surface of transport units to provide a warning that the contents of the units are dangerous substances or wastes and present risks. Transport units comprise road tank and freight vehicles, railway tank and freight wagons, and multimodal tank, tank-containers and freight containers.

Transport units carrying dangerous goods or wastes should display placards clearly visible on at least two opposing sides of the units and in such a position as to be seen by all those involved in the loading or unloading process. Where the transport unit has a multiple compartment tank which is carrying more than one dangerous substance or waste, appropriate placards should be displayed along each side at the position of the relevant compartment.

Each hazard warning placard shall be an orange rectangular panel with the UN-serial number on the bottom and the hazard warning in digits corresponding to the class of the relevant dangerous substance or waste on top.

There is a requirement for all placards and orange panels to be removed from the transport unit, or masked, as soon as the dangerous goods or their residues are discharged.

In Belgium, the FRG, France, Italy and the Netherlands, the above mentioned placarding system of the UN-Recommendations and the specific RID/
ADR-requirements are mandatory. This also applies to trans-frontier
shipment between UK and Continental Europe.

The United Kingdom has a different hazard identification placarding
system - the Hazchem-System - only for interior traffic. The orange
placard (panel) carries, in addition to the substance identification
number (UN-number), the name of the substance and the hazard warning
diamond which in itself carries the class number. Also shown is a telephone number from which specialist advice may be sought, and for use
of the emergency services, and prominently displayed is an emergency
action code, known as this Hazchem-code, which informs the emergency
services as to the medium to be employed in extinguishing any fire and
the protective clothing, evacuation etc. required by persons dealing with
the emergency.

This system is especially designed for the period immediately following
an accident until full knowledge of emergency actions pertinent to the
involved substances are available.

Regarding the "3rd Law, modifying the "Waste Disposal Law" (1985)" of
the FRG, a waste transport vehicle shall be equipped at the rear with a
white placard with the letter "A" (in German language the "A" is the
first letter of the word "Abfall" = waste).

Finally, it should be noted that in case of transport by road in France
the substance code number may be replaced by the word "Waste"(déchets).
This is not, however, allowed except for firms entitled to an exemption
issued by the "Ministry in Charge of Transport". However, this practice
should disappear further to the implementation of the new RTMD regulation
which allows to give a code number to each type of waste thank to the
N.O.S.-classes.

1.4. Performance tests of packaging (concerning sub-para 1.2.1.):

The design type of each packaging must be tested in accordance with procedures established by the appropriate authority.

Tests must be successfully performed on each packaging design type before such a packaging is used.   Such a type is defined by the design, size, material and thickness, manner of construction and packing, but may include various surface treatments.   It also includes packaging which differ from the design type only in their lower drop height.

Tests must be repeated after each modification in which the design, mate- rial or manner of construction of a packaging is altered.

The following tests are prescribed:

  drop test, leakproofness test, internal pressure (hydraulic) test, stacking test, and chemical resistance test.

Test procedures: For the preparation of a packaging for testing and the recommended test methods see UN-Recommendations, chapter 9 and Annex A. 5/ADR and Annex V/RID.

In the FRG, France, Italy and the Netherlands there is a certain conformi- ty between the specific international requirements and the relevant domes- tic regulations. In the United Kingdom, the testing procedures are not officially regulated, but industrial activities are under way to bring them in line with a transition period ending in 1990.

In Belgium, the tests on packaging are performed according to the UN- Recommendations. The tests may be used for plastic drums containing a maximum of 60 l being used for transportation of dangerous goods of ADR Classes 3, 6.1 and 8.

In Denmark, the tests are only mentioned without any reference to domestic decrees according to RID, ADR and IMDG-Code. In transit, the requirements contained in RID, ADR and IMDG-Code (Annex I) apply. Beyond the requirements for chemical waste cards, there are generally no directives for packaging of chemical wastes.

In Ireland, no performance tests of packaging for toxic and dangerous wastes are specified in national regulations. The ADR requirements apply in regard to packaging of the 25 scheduled substances under the Road Transport Regulations.

1.5. Performance tests for IBCs

Regulations for these container types will be adopted by the UN-Committee of Experts and in the RID/ADR Joint Meeting. They will become effective in 1986/87. For the time being, in international regulations, detailed requirements have not been approved.

1.6. Performance test for (freight) containers

It consists of a lifting test, stacking test, top (roof) load test, bottom load test, front or rear wall test, (distortion) rigidity test, and side wall test.

Test procedures: These performance tests are in conformity with the International Organisation for Standardisation (ISO Standards 1496). The performance test is a part of the "International Convention for Safe Containers" (CSC).

Belgium has adopted the "CSC" in the relevant "20 August 1981 Law" Denmark made references to the relevant IMO and RID/ADR-requirements. The FRG enforced the "CSC" on February 20, 1976 in the "Law on the Agreement for Safe Containers".

At the present time in <u>France</u>, only international conventions, covering containers in general, are taken into account. The RTMD makes it compulsory for containers to be authorized according to the "CSC" convention, or, where this does not apply, according to form UIC 590. Their authorization is issued by the "Containers Bureau of the SNCF".

In the RMF, an Annex C covering special requirements for containers is being developed. In the interim, containers with a "CSC" certificate are recognised for the transport of dangerous goods (circular number 2417-0.80/SN 1 of the Merchant Navy, May 16, 1980).

No performance tests for containers of toxic and dangerous wastes in <u>Ireland</u> are specified in national legislation. The National Road Transport Regulations do not specify any container performance tests for the transport of the 25 scheduled substances, but "packaging" under national regulations is intended to include "containers" so that the ADR-requirements on containers would apply to the scheduled substances.

<u>Italy</u> implemented the "Law for Safe Containers (December 2, 1972)" with the enactment of the Domestic Law 67 of February 3, 1979.

<u>The Netherlands</u> explained the test performances of containers as adopted not mentioning "CSC", but in compliance with RID, ADR and the IMDG-Code.

<u>United Kingdom</u>: The tests must comply with the relevant British Standards (B.S. 3951), which are identical with the specific ISO requirements.

1.7. <u>Performance tests for tank-containers</u>

The approval, testing and marking of the test-containers are regulated by IMO:

> IMO-type I
> IMO-type II
> IMO-type IV
> IMO-type V

and by RID/Annex X and ADR/Annex B.1b (to be published in III/1985 and in effect at the end of 1985).

In Belgium, the FRG, France, Denmark, Ireland (only RID and IMO), Italy and the Netherlands the above mentioned requirements apply.

In the United Kingdom, IMO tank-containers are tested according to British Standards (BS 3951), tank-containers for use on roads according to the "Dangerous Substances (Conveyance by Road in Road Tankers and Tank-Containers) Regulations 1981" (with some small differences). The British Rail Board in its requirements for the "design, construction, test and use of tank wagons running on BR lines", specifies modified performance tests.

1.8. Special tankcars and attached tanks

The vacuum-pressure (scavenging)-tank is a special tank, important and necessary for the treatment and carriage of some toxic and dangerous wastes.

At the RID/ADR Joint Meeting (1984) is was decided, that any substance or waste, for which a calculated pressure of not more than 4 bar is prescribed, shall be accepted for carriage in or manipulation with vacuum-pressure tanks. The "ADR-Group of Experts" deals with this problem. A draft regarding construction, equipment and use (such as mixed loading of wastes presenting different hazards in the tank) is in preparation. An amendment shall be incorporated into the Annex B.1a/RID and take into account the classification of wastes. This draft will probably come into force by the end of 1987.

1.9. Special containers for transportation of wastes

With their experience of tanks made of reinforced plastic (Annex B.1c/ADR) for transport of dangerous wastes Belgium distinguishes between this tank type and special tanks made of metal. Their design, construction, etc. follow the relevant requirements of Annexes B. 1c or B. 1a of ADR, but there are a few specific features concerning the waste problems.

Denmark uses special tankcars designed for chemical wastes.

The FRG are using only a few special tankcar types in domestic traffic
(approved design and construction), but mainly under the requirements of
Annexes B. la or B. lc of ADR.

In France, tanker-vehicles are subject to specific requirements on waste
only when they are intended for transporting waste of Classes 3, 6.1 and
8. This point is covered in the study of the collection and transportation of special wastes (July 1984) *). Here, it may be noted that within
the context of the RMF, road vehicles may be equipped with the IMO-tank-
types I, II, IV or V.

Furthermore, this study gives useful and instructive explanations which
are important for the different use of these special tanks or tank-cars.
They are commonly used world-wide e.g. in the chemical industry with
their various products. It is self-evident that they should be designed,
constructed, and tested under the specific requirements for the conveyance of hazardous substances.

It is pointed out, that tanker-wagons are subject to requirements very
similar to those imposed on tanker-vehicles or tank-containers, for
example:

- Anti-deflagration devices on ventilation openings and an earthing
  terminal for tankers intended for transporting liquids with a flashpoint lower than 50°C.

- Presence of a manhole.

- Ban on openings below level of the liquid for tankers intended to
  transport certain products (notably corrosive or toxic ones), and authorization (except for certain products) of a hand-sized hole closed by
  a flange on the lower surface of the tank.

- In case of low-level shuttering, an internal stop-cock is necessary,
  except for certain crystallisable, viscous or deep frozen substances.

---

*) "Etude sur la collecte et le transport des déchets spéciaux", by
   Mr. P. Vincent. This study had been ordered by the Franch Secre-
   tariat d'Etat à L'Environnement.

1.10. Special pressure-vacuum-tanks e.g. for road and rail

Most of the EEC Member States involved in collection, transport and treatment of hazardous wastes have such pressure-vacuum-tanks in use, but there are only domestic regulations for construction, testing and approval. For a trans-frontier shipment, international agreements are necessary.

At present, the "ADR-Group of Experts" is studying this problem and will provide a uniform solution. These specific requirements should be included into ADR and enforced before 1987.

Regulations should include: design, construction, equipment, performance testing, marking, operating procedures and special requirements.

2. Cleaning of packaging

In the different classes of RID/ADR cleaning is prescribed. The shipment prescriptions of any class deals with this problem. Depending on the properties of the dangerous substances, this manipulation shall be done with water and/or steam or special solvents. The requirements of the IMDG-Code are similar.

Apart from FRG and the Netherlands no specific provisions in national legislation or references to international agreements in the other countries exist. Denmark has a Ministerial Order (1983) for the Protection of Workers.

The measures for cleaning containers and tanks are generally precribed in marg. 10 413 (Cleaning before loading) and in marg. 10 415 (Cleaning after unloading) of the ADR regulations.

3. Applicability of special test procedures by container
    manufacturers for transport containers

Especially in case of packaging (sub-para 1.2.1.) the manufacturers are responsible for these procedures which can be always controlled by competent authorities.

In Annex A. 5/ADR (item 3550) and Annex V/RID (item 1550) it is ruled
that the design type of each packaging must be tested and approved
by the competent authority or by an authorized institution, or
by an authorized manufacturer.
The appropriate authority can require at any time that serially produced
packaging meet the requirements of the design type.

4. Relationship between international conventions and national law

Belgium

Apart from the above mentioned packaging performance tests of receptables,
the Belgian authorities have interpreted two specific types of packaging.
One is made of plastic, has a capacity of 60 l and may be used for inland
transportation of inflammable liquids, toxic and corrosive substances.
There are additional tests concerning densities and vapour pressure at
50°C, etc. The authorized testing institution is the "Belgian Packaging
Institute".
Another packaging concerns steelplate containers with a capactiy of 40 l
and is intended to be used for flammable liquids of which the vapour
pressure at 50°C does not exceed 1,1 bar.

In Denmark ADR and RID / RID-A (Danish edition) are examples of the relationship between national and international regulations and legislation.

The Federal Republic of Germany has adopted as mentioned the international
agreements. The pendants to RID/ADR are GGVE/GGVS.

France:

In general terms, containers and tank-containers receiving authorization
at a national level are also recognized for the purposes of similar international regulations (IMDG-Code or RID/ADR). Exceptions are mostly applicable in the area of   land transport and apply more to differences in
the classification of substances to be transported than to the design of
containers themselves.

It may be similarly pointed out, that control bodies such as the Bureau Véritas, Lloyd's Register or the American Bureau of Shipping, and authorizing bodies such as the Containers Bureau and the Ministry in Charge of Marine Affairs, are recognized at national level and are also authorized to carry out controls or to issue the authorizations required by international regulations.

In the area of packaging, although the ADR/RID regulations impose requirements similar to those of the RTMD, the lack of details on how tests are to be carried out leads to differences of interpretation in some countries. This is particularly noticeable in the matter of the stacking test, which in France is carried out under a non-guided load, whereas in other countries a guided load is used. This brings more stringent constraints on resistance to load for packaging licensed in France, particularly for packaging made of plastics.

Ireland:
The ADR Agreement is not explixitly implemented in national law with regard to tanks or containers used for toxic and dangerous wastes. The basic problem is the restriction of the Road Transport Regulations to 25 substances which exclude most types of dangerous wastes.

Italy:
The Italian legislation conforms to a certain extent with the specific international agreements for the transport of dangeorus goods, however, temporary rules for currently employed packaging are in force.

The Netherlands:
Summarizing the aspects concerning packaging it is concluded that Dutch regulations differ only slightly from international regulations.

United Kingdom:
The UK tank-container specification is more qualitative, compared to the quantitative approach of ADR, i.e., no values are given for such parameters as test pressures, except that the test pressure and working

pressure for any vessel operating above or below atmospheric pressure must be stated. Tank-containers for dangerous substances require a corrosion-resistant plate bearing specified particulars under ADR, only those consisting of pressure vessels need to be so equipped under the tanker regulations. There are further requirements under ADR for the various classes of danger, whereas under UK legislation all tankers which meet the general requirements may be used for any hazardous substance.

All new tanks must have type approval under ADR; it is sufficient that the manufacturer attests that the regulations are complied with within the UK.

The tanker regulations also apply to tanks permanently fixed to road tankers as they do to tank-containers, with the exception that a road tanker of any size is subject to the regulations, with similar derogations from ADR to those above.

There are substantial differences between RID and British Rail's Requirements for the design, construction, test and use of tank wagons running on BR lines.

Perhaps the most important difference between UK requirements and the international conventions is the use in the UK of the Hazchem action code rather than the Kemler number.

III. MEANS OF TRANSPORT (road, rail, sea-going and inland waterway
      ─────────────────   vessels)

1. Road vehicles

1.1. General requirements

The general provisions of the ADR regulations apply to the carriage
of dangerous substances or wastes by road in whatever quantity. Controls
over packaging, and particularly in the case of wastes are regulated and
will be achieved by the proposed classification, packaging and labelling
measures already mentioned.

The conveyance of these hazardous materials are permanently controlled by
police in some countries.

Other key factors are for instance construction of vehicles and freight
containers, special equipment and adequate inspection.

Belgium, the FRG, France, Italy and the Netherlands have a certain conformity in their domestic legislation to the ADR regulations and the above
mentioned introductions.

According to Ministerial Order No. 2 (January 2, 1985) of the Danish
Ministry of Justice tankcars and vehicles with removable tanks or container batteries must, before being put into use, be approved by the Danish
National Inspectorate of Fire Service and be marked in accordance with
the technical directives prepared by the same Inspectorate.

In case of lorry trains the hauling vehicle must be approved as well.

General requirements regarding road vehicles in Ireland are contained in
the Road Traffic Act of 1961 and the Road Traffic (Construction, Equipment
and Use of Vehicles, 1963). In addition, Ireland has implemented by regulations all the "Motor Vehicle (Type approval) Directives", adopted by the EEC.

The United Kingdom has general regulations for the commercial operation of goods vehicles, their different weights and the operators licensing (3 types: Restricted, Standard and Standard (international). Furthermore, some terms as licence holder, traffic areas, licensing authorities, certificate of professional competence and waste disposal collector are a part of regulations as well.

Other rules of the "Dangerous Substances (Conveyance by Road Tankers and Tank-Containers) Regulations 1981" concern the duty of the operator of a vehicle carrying a tank of over 3000 l or a road tanker of any size to ensure that the driver has received adequate instruction on the dangers of the substance being carried and on emergency actions, etc.

1.2. Types of road vehicles

The transport of dangerous substances or wastes by road covers different types of road vehicles. Beside the typical waste transporters like hopper- and tipper-vehicles, road tankers are used which have a tank structurally attached to, or an integral part of the frame of the vehicle. Others are tank-cars with attached tanks, which can be lifted and vehicles carrying small tank-containers, drums etc., whose capacity is 200 l and more. There is also the possibility of having packages of less than 200 l capacity.

1.3. Special requirements

The recent decisions of the RID/ADR Joint Meeting (March 1985) for the new ADR are resulting into some modifications. With regard to this, an important synopsis into the beginning of Part I of Annex B/ADR was included. The table of marg. 10 011 shows namely the limited quantities of dangerous goods in packaging, which can be transported in one conveyance unit without applicability of the following requirements (see marg. 10 010 - 10 013):

- types of vehicles (marg. XX204/Part I and II and marg. 11 205 and
  11 206 of Part II (Class I))
- driver's crew (marg. XX311/Part I and II)

- conveyance of persons (passenger) (marg. 10 325)
- Transport Emergency Cards (marg. 10 381 (1)b, 10 385 and 61 385)
- certificate of special permission for vehicles (marg. 10 282 and 11 282)
- special driver training (marg. 10 315)
- special requirements for vehicles and their equipment (all sectors 2 of Part I and II); regulations of marg. 21 212 apply
- places of loading and unloading (marg. 11 407, 21 402 and 61 407)
- operating of the vehicles (all sectors 5 (marg. 10 500 and 21 500) of Part I and II); regulations of marg. 61 515 apply.

With regard to marg. 10 010 of Part I/Annex B all transportation which is regulated according to the marg. 2201a, 2301a, 2401a, 2431a, 2471a, 2501a, 2601a, and 2801a as well as 2551a and 2651a, will be released from these requirements.

These requirements are mandatory for trans-frontier shipment.

According to § 7/GGVS and in regard to marg. 10 011/ADR the FRG is more restrictive (see Annex B. 8/GGVS with the so-called "Listenstoffe" = list of substances).

There are no special requirements applicable to the transport of wastes.

1.3.1. Conveyance in receptacles, whether on pallets or not

The transport of dangerous wastes is carried out rarely in small receptacles. It normally depends on the profitability and on the quality of the wastes. A well known mixture or solution on the way for recycling e.g. sodium cyanide wastes could be an example. In this case the packaging is a 220 l (55 gallon) drum - sometimes on pallets - and the shipper and operator must have the knowledge of the relevant regulations, especially the need for segregation.

1.3.2. Conveyance in containers other than tanks and tank-containers

When containers are used for collection and loading of dangerous substances or wastes they can require increased security, e.g. sodium cyanide wastes.

In comparison with small receptacles the "Metallic IBCs" guarantee more security for carriage of these hazardous materials; but only the double-walled types can offer sufficient security.

1.3.3. Conveyance in tank-containers and tanks

Concerning design, construction, inspection and, last but not least, operation, tank-containers and tanks are the best means of dangerous waste transport.

Depending on their use, they are designed and constructed for resistance of inside or outside effects. There are types which withstand 4 bar, 10 bar or 15 bar (overpressure) or/and which must be resistant to chemical attacks. Dangerous substances of Class 6.1 for instance with the "degree of danger" = (a)/RID must be transported in tanks with a minimum overpressure of 15 bar, or those of (b) or (c), depending on other criteria, in tanks with 10 bar or 4 bar. Tanks for transport of solids must have a minimum shell thickness of 6 mm.

1.4. Electrical devices and systems

The requirements of Annex B.2/ADR for the electrical equipment of vehicles are mandatory only for (see marg. 10 251/ADR):

a) tankers and vehicles with attached tanks including road tractors
   or such vehicles, which are transporting liquids with a flashpoint
   not exceeding 55°C or inflammable gases according to marg. 220 002;

b) vehicles transporting explosives (Class 1a, 1b and 1c) with regard to
   the requirements of marg. 11 205 (2) c).

Furthermore:

- the circuit shall be heavily insulated and be independant of the chassis,
- the wiring shall be so fixed and protected as to reduce as far as practicable any risk of damage,
- the battery shall be in an easily accessible position,
- the means of cutting off the current must be close to the battery and be effected with a double pole switch or other suitable method which shall be in an easily accessible position; and
- in any case where the road tanker is required to be provided with a fire resisting shield, the generator, battery switches and fuses shall be carried in front of that shield.

1.5. Fire extinguishing equipment

In the recent ADR Agreement, Annex B., marg. 10 240, the following measures are prescribed:

- fire extinguishing equipment needs at least two devices, one for fires in the engine compartment and one for fires in the carrying compartments. The equipment must be adequate to fight any expected type of fire, especially regarding the nature of the products to be extinguished.   The capacity of the equipment required depends on the load capacity of the vehicle;

- these fire extinguishers must not have the capacity to develop toxic gases;

- a detached parking trailer of a vehicle on a public street or place must have a portable fire extinguisher;

- details regarding the maintenance of the fire extinguisher and the drivers training for its use are outlined.

Most of the EEC Member States will accept or have accepted this modification to the former edition. However, Ireland and UK have established their own requirements.

1.6. Special safety equipment (concerning technical regulations and personnel instructions)

Every vehicle must carry among other items, 2 warning lamps ensuring that they cannot ignite the transported goods (marg. 10 260/ADR).

As discussed in Part II of this study, vehicles carrying dangerous goods must be labelled (see sub-para 1.3.2.) and equipped with an orange placard (see sub-para 1.3.4.) (marg. 10 500/ADR).

Tank units with a volume of more than 3000 l carrying dangerous substances listed in Annex B. 5/ADR must be equipped with the orange placards on each side of the tank.

In the GGVS of the FRG, the volume has been reduced to a 1000 l. Beyond this level the GGVS has prescribed limited quantities with regard to normal placarding. This concerns the Classes 1a, 1b, 1c and 6.2 with 50 kg and the Classes 2, 3, 4.1, 4.2, 4.3, 5.1, 5.2, 6.1, 8 and 9 with 1000 kg. Furthermore, this requirement is mandatory for substances in the list in § 7 of Annex B. 8/GGVS and for carrying hazardous materials in tanks and for uncleaned tanks.

The supervision of transport units with hazardous materials during short- or long-time parking is mandatory (see marg. 10 321/ADR).

1.7. Cooling and ventilation equipment

In view of the hazards of spontaneous combustion or heating of any article to be loaded on a motor vehicle, such an article shall be so loaded as to afford sufficient ventilation of the load to provide reasonable assurance against fire from this cause; in such cases the vehicle shall be unloaded as soon as practicable after reaching its destination. Marg. 21 212/ADR is concerned in this case with Class 2 goods (gases) and marg. 52 248/ADR with the cooling during transportation of Class 5.2 goods (organic peroxides).

1.8. Technical inspection

1.8.1. Special technical conditions and considerations for inspection of containers used for the transport of dangerous wastes or goods

The ADR regulations will apply to technical inspection of tank-containers, tanks and tank vehicles (IBCs from 1987 on). This will be mandatory for transit und trans-frontier shipments. Apart from Ireland, the relevant domestic regulations in the individual EEC-States are comparable with the specific ADR requirements.

The following list of organisations which can perform the required tests, approvals and/or inspections based on these regulations is provided solely as a guide:

Belgium:

Belgian Packaging Institute
Rue Picard, 15, B-1020 Bruxelles
(Approvals of test reports from this Institute are required from the "Ministerial Commission on the Transport of Dangerous Goods")

Denmark:

Dantest
Amager Boulevard 115, DK-2300 København S
Emballage- og Transportinstituttet (E.T.I.)
Meterbuen 15, DK-2740 Skovlunde

Federal Republic of Germany:

Bundesanstalt für Materialprüfung (BAM)
Unter den Eichen 87, D-1000 Berlin 45

Bundesbahn Zentralamt (BZA)
Pionierstrasse 10, D-4950 Minden

Technischer Überwachungs-Verein (TÜV) Baden, Mannheim

Technischer Überwachungs-Verein (TÜV) Bayern, München

TÜV Berlin, Berlin

TÜV Hannover, Hannover

TÜV Hessen, Eschborn/Taunus

TÜV Norddeutschland, Hamburg

TÜV Pfalz, Kaiserslautern

Rheinisch-Westfälischer TÜV, Essen

TÜV Rheinland, Köln

TÜV Saarland, Saarbrücken

TÜV Stuttgart

Germanischer Lloyd
Vorsetzen 32, D-2000 Hamburg 11

France:

Laboratoire National d'Essais (L.N.E.)
1, rue Gaston Boissier, F-75015 Paris

Bureau de Vérifications Techniques
43bis, av. de la République, F-94260 Fresnes

Italy:

Istituto Sperimentale delle Ferrovie di Stato F/S
Piazza Ippolito Nievo, 46, I-00153 Roma

The Netherlands:

Instituut TNO voor Verpakking
P.O.Box 169, NL-2600 AD Delft

TOPA Verpakking BV
10 Torenlaan, NL-2215 RW Voohout

United Kingdom:

National Testing Laboratory Accreditation Scheme (NATLAS)
Teddington, Middlesex TW11 OLW

Packaging Industries Research Association (PIRA)
Randalls Road, UK-76161 Leatherhead, Surrey

With respect to the inspection of containers, tank-containers, tanks
and vessels in several EEC Member States, six foreign classification
agencies are recognized:

Det Norske Veritas, Oslo
N.K.K., Tokio
Lloyd's Register of Shipping, London
Bureau Véritas, Paris
Germanischer Lloyd, Hamburg
Registro Italiano Navale, Genua

Furthermore, in some countries inspection by authorized insurance institutions is permissible.

The main considerations of technical inspections are mentioned in ADR
as follows:

- marg; 10 282: general explanations about testing and approval of
  vehicles, tanks and tank-containers;
- marg. 211 400: approval of tank (metal) design types;
- marg. 211 150 - 211 154: test procedures for tanks;
- marg.211 350: special requirements for tanks used for Class 3, with
  a minimum testing pressure of 4 bar (e.g. for carbon disulfide);
- marg. 212 140: approval of tank-containers design types;
- marg. 212 150 - 212 154: test procedures of tank-containers;
- marg. 212 350: special requirements for tank-containers used for
  Class 3, with a minimum testing pressure of 4 bar;
- marg. 213 141 - 213 143: testing of reinforced plastic tanks;
- marg. 214 000 - 214 285; testing of materials for tanks and tank-
  containers.

1.8.2. Special technical conditions for inspection of equipment of road vehicles used for the transport of dangerous goods

The marg. 10 282 and 230 00 - 230 001 deal with, among other things, the regulations for tank vehicles and certificates of approval.

1.8.3. Authorization of examiners and experts for road vehicles and containers

Performance testing and repetition testing in accordance with the require- ments mentioned in sub-para 1.8.1. and 1.8.2. for tanks, tank-containers and motor vehicles must be carried out by experts of technical control institutions (mentioned in sub-para 1.8.1.).

An expert is a person who

a) can warrant, on the basis of his training, knowledge and practical experience, that a test has been properly conducted,
b) has the personal qualities required to be worthy of trust, and
c) is capable, in the course of his expert activity, to apply the instructions correctly.

An examiner has similar knowledge but, due to having lesser education and training than the expert, cannot make the responsible test decision and issue licenses.

Belgium:

The inspection must be performed by a competent authority or a licensed expert and must be certified.

Denmark:

Authorization of examiners for road vehicles can be obtained by the traffic police.

Federal Republic of Germany:

Only experts, who must have an official attestation from the above mentioned test control institutions, and from the competent authorities of the German Länder, are authorized to make a mandatory testing decision and issue licenses. The approval authority is the BAM.

France:

Besides institutions authorized by bodies as the "Containers Bureau" and the "Ministry in charge of Marine Affairs" ("Direction des Ports et de la Navigation maritime") or the "Ministère de l'Urbanisme, du Logement et des Transports" there are control bodies such as the "Bureau Véritas", "Lloyd's Register" or the "American Bureau of Shipping".

Ireland:

Concerning the "Vehicles Testing Regulations of 1981" (in accordance to the EEC-Directives 43/77 and 47/77) the schedule of tests is not specific regarding vehicles for the carriage of dangerous substances, but is generally designed for heavy goods vehicles. Some 90 garages have been approved to carry out the roadworthiness tests required by the regulations.

There are no national inspection facilities for testing of vehicles specifically determined for carriage of dangerous wastes.

Italy:

In Italy inspections are made by bodies depending directly from the Ministry of Transport. For vehicles and tanks these bodies are the Provincial Inspectorates of the Civil Motorization , for packaging the above mentioned Istituto Sperimentale FS (see page 67).

The Netherlands:

For examination and testing, the following agencies are authorized:
- Dienst voor Stoomwezen (testing);
- the TNO Institute for Packing (testing);
- the KCGS (examination only during transport).

As far as the ADR allows, private agencies may be appointed to execute test regulations.

The director of the "Rijksdienst voor het Wegverkeer" (State Agency for Road Traffic) is authorized to:

- issue further regulations for the examination of vehicles (tank-containers),
- condemn vehicles (tank-containers) if the regulations are not met,
- withdraw the examination document if, after an accident, a vehicle or a tank-container is not offered for examination.

United Kingdom:

The examination must be carried out by a competent person, who has such practical and theoretical knowledge as will enable him to determine defects or weaknesses which the examination is designed to uncover and to assess their importance. If the competent person is employed by the operator, his judgement ought not to be clouded by operational requirements. For this reason, many companies prefer to use an independent examiner from outside the company.

1.9. Relationship between international conventions and national law

As mentioned above, apart from Ireland, Luxembourg, and in a few cases UK and Denmark, there is a certain relationship between the specific ADR and the domestic requirements.

2. Rail vehicles

2.1. General requirements

According to the recent developments in the international legislation of the transport of hazardous materials by rail and road relating to harmonization of laws, the new RID version became internationally legal in May 1, 1985. The transposition of these agreements to the relevant domestic legislation of the EEC Member States has already been or shall be carried out in the future.

2.2. Transport in receptacles whether on pallets or not or in containers other than tanks or tank-containers

The RID and ADR regulations are similar. Beside the application of the receptacles or containers, the consignment procedures, segregation etc. are regulated by the specific chapter 2 of every Class.

2.3. Transport in tank-containers or in railway tankcars (tank-wagons)

With regard to design, construction, inspection, and last but not least, operation, tank-containers or attached tanks are the best means for dangerous good transportation. For details regarding for instance the application of tank-containers see Annex X/RID / point 1.3.3. of this part.

2.4. Special safety equipment

The marg. 1400 - 1402, Annex IV/RID deal with the regulations for rail wagons regarding electric devices. Measures for avoidance of unexpected ignition of a possible explosive atmosphere in the wagons are given.

2.5. Technical inspection

The main positions in RID in these matters are:

Annex X mentions the regulations about design, construction, equipment, testing procedures, licensing (approval), marking and operating of tank-containers. This applies in case of some items (specific substances requirements) of every Class, depending on additional dangerous properties or special prescriptions.

In Annex XI, which mentions railway tankcars, similar regulations exist as mentioned above in reference to Annex X.

In comparison with sub-para 1.8.1./Part III there are certain conformities between the technical inspection of tanks or tank-containers and vehicles used by road or rail.

Some of the testing institutions listed in sub-para 1.8.1. are engaged in this matter in their work on transportation by rail; but several railway administrations have their own service, as follows:

Denmark:

The authorization of examiners for rail vehicles and containers lies with DSB: Danske Statsbaner (Danish State Railways) Godstjenesten, Kalvebod Brygge 34, 5, DK-1560 København V.

Containers must satisfy ISO Standard 1496 and tests mentioned in CSC.

In the FRG the BAM (see sub-para 1.8.1./Part III) is responsible for technical licensing (approval) of IBCs for tank-containers while the "Bundesbahn-Zentralamt" (see sub-para 1.8.1.) is competent for railway tankcars and vehicles as well as for the entire testing procedure.

France:

The only technical inspections are these covering the authorization of tanker-wagons. These inspections consist of checks concerning the conformity of the wagon presented with the data of the prototype. Wagons which are not precisely identical to the prototype, but which have been constructed without any modification liable to call the test results into question, are not subject to new trials.

The examinations mentioned above are carried out under the control of the "Containers Bureau of the SNCF". This Bureau is also responsible for giving the wagon a registration number.

The "Containers Bureau" is the official body responsible for authorizing, under the RTMD (Règlement pour le Transport des Matières Dangereuses), containers, tank-containers and tank-wagons. It is also entitled to issue authorizations corresponding to the CSC conventions, as well as RID/ADR regulations.

Ireland:

For the inspection and examination of   containers, tank-containers and tanks the regulations of the RID apply.

CIE (the National Rail Transport Authority) has the necessary facilities
and expertise to carry out the technical inspection required.

The Netherlands:

For the inspection and examination of   containers, tank-containers
and tanks the regulations of the RID apply.

For examination and testing the following agencies are authorized:

- Dienst voor het Stoomwezen (testing)
- the TNO Institute for Packaging (testing)
- NV Nederlandse Spoorwegen (Dutch Railroad Company)
- State Traffic Inspection (examination).

Some of the performance tests are executed under contract in the FRG and
France in case the above mentioned agencies do not have the adequate testing equipment.

Where the RID allows, private agencies may be appointed to execute test
regulations of a margin number.

United Kingdom:

All wagons must undergo an initial inspection before being put into service
by the "Mechanical and Electrical Engineering Department of the British
Railways Board". In the case of tanks and tank-wagons, this will ensure
that the predetermined specifications have been met and will include an
hydraulic pressure test to the test pressure indicated on the attached data
plate. An external inspection of each tank must be carried out at least
every 3 ½ years (RID: 4 years) and the internal examinations are normally
carried out every 7 years (RID: 8 years).

The inspection authority must be either a recognized "Pressure Vessel
Insurance Company" or, if some other body is appointed by the owners, it
must be approved by the BR Quality Assurance Engineer.

Italy:

For the inspection and examination of   containers, tank-containers
and tanks the RID regulations apply. The authorized agency is the Istituto
Sperimentale FS (see page 67).

2.6. Special technical conditions and considerations for the transport of dangerous wastes

In Denmark, the FRG, France, the Netherlands and the UK there are no requirements specifically for wastes in the domestic railway transport regulations,
but any wastes carried   must comply with the relevant RID requirements, and
with the consignment procedures in the special national and international
wastes regulations, inclusive the permissions.

2.7. Authorization of examiners and experts for rail vehicles and containers

This matter is discussed in detail in the relevant explanations of road
transport.

2.8. Relationship between international conventions and national law

There are no major differences between domestic regulations of the EEC
Member States and the RID.

3. Transport by water

3.1. Inland water vessels

3.1.1. General requirements

Federal Republic of Germany: Design, construction, equipment, inspection,
operation, etc. of the inland waterway vessels are in accordance with ADNR.
The domestic regulation is the Gefahr-Gut-Verordnung-Binnen-Schiffahrt
(GGVBinSch).

The transport of dangerous wastes by inland waterway vessels is strictly
controlled by the specific waste laws.

Nevertheless, this type of transportation is on the decline since it is only
used to carry wastes to seaports and to trans-ship to a marine vessel
which dumps the waste in the open sea. According to the "Convention on the

Marine Pollution by Dumping of Wastes and other Matters" (MARPOL 73), the FRG will discontinue these operations during the nineteeneighties.

France:

The current requirements are to be replaced in the near future, as regards to construction of ships, by the future Appendix 7 of RTMD, which is very close to ADN. Furthermore, the report generally mentioned several requirements of the ADN, especially with reference to the properties of the dangerous goods in respect to the safety of the vessel.

The only authorizations necessary for the transport of dangerous wastes are the certificates of entry into service required for any tanker-ship intended for the transport of dangerous substances.

Leaving aside any difficulties with classification in the nomenclature, the transport of waste in containers poses no special problems on the level of transport in the strict sense. However, like any lengthy transportation process, it is not well suited to sludgy waste products whose solid components tend to decant and thicken. In practice, it would appear that in France inland waterway transport is only used for some non-dangerous solid waste products (gypsum phosphates, household waste).

United Kingdom:

For waterways regulated by the "British Waterways Board", the movement of dangerous goods is controlled by the terms and conditions for the "Transport of Dangerous Goods on the Board's Waterways and Docks". For estuaries and other larger navigable waterways, the pending legislation is the "Dangerous Substances in Harbours and Harbour Areas Regulations (1985)".

The most common movements of hazardous goods in barges in the UK are the removal for dumping at sea of various industrial and domestic sewages, but this must be licensed under the provisions of the "Dumping at Sea Act (1974)". For such cargos no receptacle is needed the substance merely being dropped into the bottom of specially constructed vessels.
In view of the fact that there are few movements of dangerous wastes on UK waterways, there has been little opportunity for technical difficulties to manifest themselves.

Italy:

Design, construction, equipment, inspection, operation etc. of the inland
waterway vessels are controlled by the Provincial Inspectorates of the Civil
Motorization.

3.1.2. Authorization of examiners and experts for vessels and containers

In the FRG the experts of the "Germanischer Lloyd" (see sub-para 1.8.1.)
and of the "Inland Waterways Transport Injuries Insurance Institutes" are
authorized to perform the technical inspection. For approval of tank-
containers, the BAM is responsible.

In France the inspection of containers is carried out under the supervision of the "Containers Bureau", in compliance with the RTMD.

The inspection of boats is carried out by an expert chosen by the owners
and recognized by the "Ministry in Charge of Transport". Generally, the
expert agency is a company carrying out classifications on an international
scale.

In the Netherlands, for the testing of packaging the regulations of ADNR
apply. This means that the agencies mentioned in sub-para 1.8.1. are also
authorized for the inspection of packaging for the transport of inland
waterways.
The inspection on vessels is executed by the KCGS.
For Italy see page 70.

In UK barges carrying wastes have no specific technical requirements:
there are no authorized examiners specifically for such vessels.

3.2. Sea-going vessels

3.2.1. General requirements

In the EEC Member States the IMDG-Code applies to the transport of dangerous goods, that means the specific requirements concerning colis, containers, tank-containers and tanks are regulated under this Code. There is a
conformity between the UN-Recommendations and the IMDG-Code and a harmonizing development with RID/ADR on the basis of the UN-Recommendations.

In comparison with the IMDG-Code there are some differences or additions
in domestic regulations:

Belgium:

The regulations regarding the transportation of dangerous goods at sea
are part of the Belgian "Maritime Inspection Code". In this Code several
requirements have been established regarding the safety of the ships and
the certificates required. Ships intended for bulk transport of dangerous goods must have a special certificate.

Denmark:

Directives are contained in Part VII of notifications "B" of the Danish
Ships Inspection Service for the construction and equipment of ships, etc.

The transport in receptacles or in containers must be executed in accordance with the IMDG-Code or for the western part of the Baltic Sea according to the current "Memorandum of Understanding" (edition 1985).

Federal Republic of Germany:

Apart from the IMDG-Code the "Memorandum of Understanding" is mandatory.
For the transport of receptacles the Annex "RM 001" of the GGVSee applies.
For the approval of the latter the BAM is competent.

France:

The French regulations have no specific requirements for the construction
and layout of ships intended for the transport of dangerous substances
or wastes.

The only regulations followed by shipowners are the international regulations (SOLAS 1974 , MARPOL 1973, Protocol 1978). In any case, all French
ships must be authorized by the "Ministry in Charge of Maritime Affairs".

Ireland:

The relevant IMDG-Code concerning the transport of dangerous goods applies.
The "Merchant Shipping Act (Dangerous Goods) Rules 1983" (S.I. No. 306/1983)

contains the key regulatory requirements in this area. It seems that the definition of "goods" in this context is broad enough to include "wastes". The 1983 rules apply to ships registered in Ireland and to other ships while they are loading or unloading a cargo within the country.

United Kingdom:

The "Merchant Shipping (Dangerous Goods) Regulations 1981" (S.I. 1981/1747) correspond to chapter VII (Carriage of Dangerous Goods) of SOLAS 1974, which was ratified by the UK in November 1979 .

3.2.2. Water-borne transport of dangerous goods

Denmark:

Dangerous waste is packed, transported and classified as registered cargo corresponding to the dangerous types of cargo described in the IMDG-Code.

Federal Republic of Germany:

The transport of hazardous wastes applies to dumping activity of sea-going vessels which is strictly controlled. This activity will be stopped in the 1980s.   There are only a few small ships involved in this activity.

France:

Like the inland waterway transport, sea transport poses no special problems with regard to the transport of waste, apart from possible difficulties in the classification of substances, and from the fact that this transport is not well suited, for long voyage, to sludgy wastes in which solid compo- nents decant.

Apart from the transport of waste vehicles, which operate in a regular ferry service between continental France and its islands (e.g. Corsica), there is only one case of waste transport by sea, that is incineration at sea. This procedure is generally executed by foreign ships.

Italy: The rules concerning the transport of dangerous goods by sea are the application ordinances of the DFR No 1008 which follows the IMO rules. For the approval of these transports a certificate of the Italian Naval Register (RINA) is requested, followed by an approval of the Ministry of Merchant Navy.

3.2.3. Technical inspection

In Denmark the "Port State Control", a division of the "Danish Government Ships Inspection Service", occasionally carries out inspections concerning transport of dangerous goods.

In Italy the harbour-offices, which depend on the Ministry of Merchant Navy, make occasional inspections regarding the transport of dangerous goods.

In the FRG the competent authorities are the Ministry of Transport, the specific Port administrations, the BAM, the "Germanischer Lloyd" and the "Mercantile Maritime Industrial Injuries Insurance Institute". They are mandatory for the design, construction, testing, approval and conditioning of ships and packaging regarding to the GGVSee. The Ministries of Interior Affairs of the concerned Länder are responsible with waste transport.

In the Netherlands the regulations of the IMDG-Code apply to the construction and equipment of sea-going vessels. Additional rules are given in chapter E, Annex IV to the Ships Decree including general regulations for the construction, especially with regard to fire-extinguishing equipment.

3.2.4. Authorization of examiners and experts for vessels and packaging (tanks etc.)

In Denmark the Port State Control carries out this service.

In the FRG the experts of the institutions mentioned under sub-para 3.2.3 (see above) are authorized.

In Italy, experts of the Italian Naval Register (RINA) carry out this duty depending on the harbour-offices.

In the Netherlands the Shipping Inspection is in charge of the supervision of the Ships Decree, e.g. the controllers of this inspection supervise the fulfillment of the regulations with respect to the transport of dangerous goods to and from sea. They are authorized:

- to ensure that the regulations are executed;
- to check all the goods mentioned in the Ships Decree;
- to seize a ship if the regulations are not met.

With respect to the inspection of containers, tank-containers, tanks
and vessels, seven foreign classification agencies, as mentioned under
sub-para 1.8.1., are recognized.

Bureau Veritas, Germanischer Lloyd and Lloyd's Register of Shipping have
also been appointed for the execution of the margin numbers of the ADNR
in the cases where a classification office is allowed or required.

3.3. Relationship between international conventions and national law

Apart from Denmark, Ireland and Luxembourg, all the EEC Member States are
signatories to the "Regulations for the Transport of Dangerous Goods on
the Rhine" (ADNR) and apart from Luxembourg signatories to the IMDG-Code.

4. Technical problems and existing regulations concerning combined traffic

The so-called "Piggy-back transport" is a method of conveyance (e.g. a
truck on a rail-lorry) by different means of transport. In comparison with
North America, Western Europe has different technical systems, like "Rail-
Route", "Flexi-Van", "Semi-Remorque" etc. Design and construction are
being constantly developed. The aim is to achieve a "Rolling Highway"-
system. In case of the transport of dangerous wastes these systems are
more or less unused.

In France a single set of regulations deals with road, rail and river transport. As a result, there are no problems with the regulations covering
combined transport, as waste is classified on a single scale and the authorization of the various equipments used for transport is issued by the same
authority. It should, however, be noted that combined transport over land
is not used in France for the transportation of dangerous waste.

On the other hand, combined road-sea transport is regularly used on routes
such as Corsica and the continent. Here again, there are no special

technical or regulatory problems. The RMF authorizes the transport of tank-vehicles and tank-wagons recognized by the RTMD for "short international voyages".

Combined road-sea-transport applies to "short international voyages" during which the ship does not go more than 200 miles from any port or from any place where the crew and the passengers can be brought to safety. The distance between the last port of call (where the voyage starts) and the port of final destination does not exceed 600 miles.

The packaging, labelling, orange colour placarding or segregation requirements are regulated according the RID/ADR and not to the IMDG-Code (there are different criteria and classification aspects depending on whether the hazardous materials are transported by sea or land).

With regard to this problem, Denmark, Sweden, Finland and the FRG concluded an agreement in 1980, titled "Memorandum on the transport of dangerous substances in the Baltic Sea", regulating these sea-transports in the Western Baltic Sea. The newest edition came into force in 1985. The IMO has been consulted with these problems. The consultations are now in the drafting stage and the new requirements will be included in the IMDG-Code.

According to European arrangements, it will apply to the "short international voyages" accross the English Channel, the Irish Sea or in the Mediterranean Sea between Corsica, Sardinia and Sicily and the Continent.

Moreover, the RMF (France) also recognizes the validity of the labelling requirements (apart from exemptions) required by the overland transport regulations for road vehicles and wagons, whatever the type of sea-voyage (long or short) may be.

This set of adaptations of the RMF means that combined road-sea transport, used only in the transport of waste, can be carried out without any regulatory difficulties.

One point of divergence deserves mention: depression equipment on tankers, obligatory under the marine regulations, is authorized only according to the land regulations. This means that a certain number of sealed containers, currently used for land transport, may not be used for sea transport.

IV. LOADING AND UNLOADING

Apart from Ireland the other EEC Member States have ratified the ADR agreement and transposed it into their domestic legislation. Thus, they apply the relevant requirements of loading and unloading as mentioned below. In Ireland there are only general provisions in this area.

On the other hand, there are specific port regulations or relevant legislation concerning industrial law in all countries.

1. General requirements

If dangerous goods or wastes are involved in loading and unloading procedures, there are some basic security requirements which apply to most of the different Classes of ADR:

a) No smoking while loading and unloading: Smoking in or around any motor vehicle is forbidden while manipulating any explosive, flammable liquid or solid oxidizing material, or flammable compressed gas.

b) Keep fire and other ignition sources away during loading and unloading proceedings: This applies to all hazardous materials listed under sub-para I a), which could be sources of fire and explosion.

c) Set handbrake while loading and unloading: No hazardous materials shall be loaded in or on, or unloaded from any motor vehicle unless the handbrake and the brakeblocks are securely set and all other reasonable precautions have been taken to prevent motion of the motor vehicle during such a loading and unloading process. The breakage of the filling hose, following an unexpected movement of the vehicle, can give rise to a spillage and the development of an explosive atmosphere.

d) Secure packages in vehicle: Any tank, drum, gas cylinder, or other packaging, not permanently attached to a motor vehicle, which contains any dangerous goods, must be secured against movement within the vehicle on which it is being transported.

e) <u>Prevent relative motion between containers</u>: Containers of dangerous liquids and gases must be so braced as to prevent motion relative to the vehicle while in transit. Containers having valves or other fittings must be so loaded that there will be the minimum likelihood of damage during transportation.

f) <u>Take precautions concerning containers in transit or other long distance movements</u>: Reasonable care should be taken to prevent undue rise in temperature of containers and their contents during the journey.

g) <u>Attendance requirements</u>:

1) <u>Loading</u>: A cargo tank must be attended by a qualified person at all times when it is being loaded. The person who is responsible for loading the cargo tank is also responsible for ensuring that it is so attended.

2) <u>Unloading</u>: A motor carrier who transports hazardous materials by a cargo tank must ensure that the cargo tank is attended by a qualified person at all times during unloading. However, the carrier's obligation to ensure attendance during unloading ceases when

- the <u>carrier's obligation</u> for transporting the materials is fulfilled;
- the <u>cargo tank</u> has been placed in the consignee's premises; and
- the <u>motive power</u> has been <u>removed</u> from the premises.

3) <u>A person is "qualified"</u>, if he has been made aware of the nature of the hazardous material which is to be loaded and unloaded and has been instructed on the procedure to be followed in emergencies.

4) A <u>delivery hose</u> or <u>flexible pipe</u>, when attached to the cargo tank, is considered as being a <u>part of the vehicle</u>.

5) It should be ensured that the <u>rubber hose and flexible metallic pipes</u>, used for loading and unloading procedures of dangerous liquids or gases, are well constructed, have their <u>performance</u> and <u>repetition tested</u> and have permanent <u>maintenance and monitoring</u>.

6) <u>In public places or areas</u> prohibited loading procedures are only possible with the <u>permission of the competent authorities</u>.

h) <u>Prohibited loading combinations</u>: In any single driven motor vehicle or any single unit of a combination of vehicles, hazardous materials shall be loaded together only if not prohibited by loading and storage charts.

k) <u>Inflammable materials for stowage</u>: It is prohibited to use highly inflammable materials for the stowage of dangerous goods to secure their movements during transport.

2. <u>Special regulations for loading and unloading</u>

The RID/ADR regulate loading and unloading, especially the filling rates and measures for

<u>tank-containers</u> in Annex X/RID, item 1.7; and in
Annex B. 1b/ADR, marg. 212 170 - 212 177 and for
<u>tanks</u> in Annex XI/RID, item 1.7; and in
Annex B. 1a/ADR, marg. 211 270 - 211 279.

3. <u>The basic requirements for loading and unloading</u>

Mentioned in Part IV, para 1. above are:

a) For Class 1a, 1c, the marg. 128 (only FRG) and 154 of RID, and marg. 11 407 and 11 414 of ADR.

The proceedings must consider the public, and unloading may only occur in certain places (if practicable only by day).

b) Marg. 10 417/ADR requires that tanks for liquids with a flashpoint below 55°C and tanks for inflammable gases must be earthed during loading and unloading; and
marg. 10 414 regulates handling and stowage in general,
marg. 10 419 loading and unloading, and
marg. 10 431 the operation of the engine during loading and unloading.

c) For Class 2 (gases), marg. 21 407/ADR regulates the point 1.g) 6) and marg. 21 414/ADR points 1.d), 1.e), 1.f) and 1.g).

d) For Class 3 (flammable liquids), marg. 31 414/ADR regulates the stowage with inflammable materials (sub-para 1.k)).

e) For Class 4.2 (substances liable to spontaneous combustion), marg. 42 414 concerns points 1.d) and 1.k).

f) For Class 4.3 (substances which, in contact with water, emit inflammable gases), marg. 43 414 corresponds with sub-para 1.d); in addition any contact with water must be avoided.

g) Class 5.1 (oxidizing substances) is regulated by marg. 51 414. The relevant rules are those of sub-paras 1.d), 1.f) and 1.k). In addition, the receptacles must be stowed so that they cannot be upset.

h) For Class 5.2 (organic peroxides), marg. 52 414/ADR provides that particular attention be paid to sub-paras 1.d), 1.f) and 1.k).

i) Class 6.1 (poisonous substances), marg. 61 407 requires that loading procedures are only possible with the permission of the competent authorities.

k) For Class 8 (corrosive substances), marg. 81 414/ADR corresponds to the sub-paras 1.d), 1.e) and 1.k).

4. Specific legislation, particular industrial laws for the protection of workers

Denmark:

With reference to the "Ministerial Order on National Road Transport of Dangerous Goods" (including chemical wastes) of January 2, 1985, loading, stowage, and unloading must be performed in accordance with the regulations in the "Technical Directives" issued by the Danish National Inspectorate of Fire Service and with the national additions to ADR.

Further regulations are: "Regulations concerning the loading and unloading of ships" (Public. 50/1075); "Instructions for Hookmen".

Federal Republic of Germany:

Specific legislation in these matters consists of:

"Regulation on the Storage, Filling and Conveyance of Combustible Liquids on Land" (Verordnung zur Lagerung, Abfüllung und Beförderung brennbarer Flüssigkeiten zu Lande) (February 27, 1980)

"Loading and unloading of vessels" (Be- und Entladen von Wasserfahrzeugen) (1978)

"Regulations of the Inland Waterways Transport Industrial Injuries Insurance Institute" (Vorschriften der Binnenschiffahrts-Berufsgenossenschaft)

"Accident Prevention Regulations of the Mercantile Marine Industrial Injuries Insurance Institute" (Unfallverhütungsvorschriften für Unternehmen der Seefahrt) (UVV/See) (1981).

Ireland:

In the area of road transport the absence of specific national regulations, and the restriction of the scheduled substances to 25, means that the ADR agreement is not effectively applied to loading and unloading of dangerous wastes. The restrictive scheduled list also applies to rail transport, though there are general provisions in the draft bye-laws on loading and unloading the scheduled dangerous substances. The most extensive requirements in relation to loading and unloading are in the area of sea transport, where relevant IMDG-Code and International Cargo Handling Co-ordination Association Code are applied.

The Netherlands:

Equipment regulations regarding loading and unloading are given in

- the Nuisance Act
- the Stevedores Act
- the Dangerous Tools Act.

In case that a license on the basis of the Nuisance Act is required for an establishment, regulations can be given for the equipment. These regulations may include technical demands.

The Stevedores Act also provides regulations for the equipment, e.g.:

- means of transport to and from workshops
- steel specifications for masts and derricks
- inspection of loading and unloading equipment, lift-trucks,
  hoists, winches etc.

The Dangerous Tools Act is a general act, which also comprises regulations for loading and unloading equipment, e.g., with regard to winches, trucks, electric powered devices, etc.

United Kingdom:

There is a general duty under the "Health and Safety at Work, etc. Act 1974" for employers to protect, as far as practicable the health and safety of any person employed by them, and any person who may be affected by their own acts or omissions or those of any of their employees on their premises in the course of their work. In this context, premises include vehicles.

In the British Rail working manual for rail staff the pink pages outline requirements for loading and unloading dangerous goods for transport by both freight and passenger trains.

The tanker regulations define conveyance by road as starting from the beginning of loading of any dangerous substance and ending with the cleaning of the tank. For this reason, all requirements relating to safety also apply during loading and unloading. The Dangerous Substances Act (Conveyance by Road in Packages etc.) will contain a similar provision.

Italy:

The requisites for the protection of workers are included in the application rules of the national laws, referring to different modes of transport: "FS Regulations" for railway transports, DFR No 1008 for transports by sea and "Highway Code" for road transports.

5. Technical regulations concerning the loading of dangerous wastes

In relation to the RID/ADR decision (Joint Meeting in March 1985), dangerous wastes will be transported as common dangerous goods. Special technical regulations concerning their loading are therefore not necessary.

V. CONTROL AND MONITORING SYSTEMS

1. Control and monitoring systems for supervising transport of dangerous wastes

According to the international conventions, control and monitoring systems for transport of dangerous wastes include only a few written instructions and transport documents. Therefore, most of the Member States have developed and implemented their own more stringent control and monitoring systems. In some cases (FRG, I, IRL, L, UK) a trip ticket system is implemented: a document ("trip ticket") must accompany the transportation of the substance. All concerned persons must complete the document, and copies of the document must be sent to the relevant competent authorities. This system ensures that all dangerous wastes will be disposed of by authorised disposers on licensed sites. The competent authorities are able to control the entire management of the dangerous wastes. The EEC Directive 84/631 aims at an uniform implementation of a similar system, and outlines a number of rules. The trip ticket systems which are already in use are not identical for every country, but still they look alike in broad outline. An example of an apparently well monitored trip ticket system is the German one: the producer has to fill in a sixfold "trip ticket" with the following description of the waste:

a) name, classification, nature and aggregation
b) quantity
c) registration number of the transport vehicle
d) name of the waste producer
e) name of the waste carrier
f) name of the disposal or the tipping facilities.

After taking over the waste the carrier presents the acknowledgement receipt to the producer as follows:

The 1st copy of the trip ticket is given to the waste producer for his record.

The 2nd copy is given to the authority responsible for the waste producer.

The other 4 copies must accompany the transport and must be handled over to the disposer, who has to acknowledge the acceptance of the waste and to distribute the other copies as follows:

copy 3 for the record of the carrier;

copy 4 for the competent authority of the disposer;

copy 5 acknowledgement of acceptance to the waste producer;

copy 6 for the record of the disposer.

In the British system, the trip ticket or consignment note must normally also be filled in sixfold. Parts A, B, C and D of it are completed before the waste is moved.

The producer must send one copy, with parts A and B filled in, to the disposal authority for the area in which the waste is to be disposed of. If a carrier is used, the carrier must fill in part C and the producer part D on the remaining 5 copies at the time of collection. The producer must retain one copy and, if the waste is to be disposed of in the area of an authority other than that in which is was produced, a copy must be sent to the producer's own authority. The other three copies are sent with the carrier. The disposer, on receiving the waste and disposing of it, must complete part E, keep one copy, give one copy to the carrier and send one copy to the disposal authority where the waste was produced. This means that the authority where the waste is produced must be notified of its disposal. Any discrepancies may then be investigated. The system in use in the UK, by making the carrier certify that he has accepted the waste and by the use of the notification procedures, although pre-dating the "trip-ticket" system described in EEC Directive 84/631 incorporates similar provisions.

In France, a form of trip-ticket system has come into force from July 1985 on : a progress schedule accompanies the waste to its destination. It must be countersigned by each party involved, all of whom keep copies, and it must be returned within the month following the shipment to the producer, with an endorsement stating that the waste has been taken in charge by the centre to which it was addressed.

Producers, collectors or transporters and managers of centres must maintain an up-to-date register of operations and send a three-monthly summary to the department in charge of checking registered installations.

The trip ticket system has not been implemented yet in Belgium, Denmark
and the Netherlands In the Netherlands the control and monitoring of the
transport legislation conforms almost entirely with the international
conventions. In Dutch environmental legislation there is a "cradle
to grave" control system regarding the transfer of chemical waste. In
Belgium a law has now come into force (as from July 9, 1984), which aims
at an implementation in the near future of a compulsory notification of
activities involving transport of dangerous waste.

The control and monitoring systems for the export of dangerous waste is to
be harmonized on the basis of the trip ticket system proposed in the EEC
Directive 84/631. The main differences between this system and those
already in use are:

- in the EEC system the holder of the waste also has to make notifications
  regarding assurances and safety in traffic;

- in the EEC system all countries involved have to be notified in case
  of export of dangerous wastes. The country of destination can make
  some motivated objections while the countries of transit and dispatch
  can lay down appropriate conditions. The holder of the waste must comply with these conditions and must solve the problems giving rise to
  the objections, in order to make the shipment;

- in the EEC system the holder of the waste may not execute the trans-
  frontier shipment before the competent authority (in the country of
  destination) has sent him an acknowledgement of the notification. The
  acknowledgement has to be entered on the consignment note.

2. <u>Control and monitoring procedures for safety in traffic, for</u>
   <u>ancillary equipment, for packaging, for loading and unloading</u>
   <u>equipment and for loads</u>

Information about the control and monitoring regarding safety in transport
of dangerous wastes can be found in Part VII (e.g. driver training and
licensing) and regarding ancillary equipment, packaging and loading and
unloading equipment in Parts II, III and IV (e.g. safety labels on packaging and vehicles, technical inspection of vehicles).

In <u>Belgium</u> there is an interesting development on the territory of safety:
accident reporting. Safety officers may be asked to report on accidents,
causing injury to employees in a stipulated form. The duty of
the officer further consists of <u>analysing the accidents and proposing</u>
<u>preventive measures</u>. With regard to road transportation an information
table has been developed for the police to use in reporting accidents on
the road with vehicles carrying dangerous goods. On the basis of this information the Institute for Road Transport makes a statistical study every
year including an evaluation of the causes of the accidents and of their
effects.

3. <u>Research development</u>

In the <u>United Kingdom</u> there are some interesting developments in the research on control and monitoring of transport of dangerous wastes, e.g.:

- the "Oregon" system: The University of Nottingham (UK) is engaged in
  research in the USA on a system of uniquely identifying a road vehicle
  by means of an electronic number plate fixed underneath. This tracking
  system is at present monitoring heavy goods vehicles at state borders and
  one or two points in between, a process made possible
  by US   law which forces vehicles above a given size to use only specified routes. The system relies on an inexpensive microwave generator on
  each vehicle. The advantages of such a system applied to dangerous cargos are obvious, and firms are encouraged to use the system so they may
  know the whereabouts of each vehicle of their fleet.

- satellite tracking, in UK already in use for shipping, could be eventually extended to include road vehicles. The advantage of such a system would be in giving the precise location of a vehicle at any time, but set against this would be the fact that to equip a road vehicle with a suitable radio transmitter would be considerably more expensive than equiping it with a simple microwave transmitting device. It is considered that one satellite of the "Navstar" type could locate every vehicle in the UK.

In chapter VI of this report other (electronic) systems will be dealt with.

VI. USE OF ELECTRONIC DATA PROCESSING SYSTEMS FOR THE ORGANISATION
     AND SUPERVISION OF THE TRANS-FRONTIER SHIPMENT OF DANGEROUS
     WASTES (SUBSTANCES)

To date achievements have been made in several countries in developing
electronic data processing systems for the organisation and supervision
of the trans-frontier shipment of dangerous wastes or substances.

In Dutch environmental legislation, computers are being used for the
storage and processing of the information received from each person/firm
as a result of the notification system. This notification system
(a "cradle-to-grave" system) enables one to decide whether the waste is
properly disposed of. The EDP system also provides an opportunity to
trace those generators suspected in failing to notify. In France a similar system has come into force since July 1985. In both systems the information about disposal of exported wastes is being or will be excluded.
Regarding transport in the Netherlands (as in a number of other countries)
the only EDP system in existence is for railway transport. This system,
the so-called "System Gegevensverwerking Goederen SGG" (System Datapro-
cessing Transported Goods), is used by the NV Nederlandse Spoorwegen
(Dutch Railroads) who register the following information for every wagon:

- station of departure and arrival
- train number in which the wagon is placed
- location of the specific wagon in train
- sender and destination (names and adresses) of the load
- name of the substances that are being transported (by means of a
  SGG-code number).

The SGG-code numbers for substances are also used for an immediate determination of the measurements which should be taken in the event of cala-
mities.

A similar system for railway transport is being used in the United Kingdom
(TOPS), Italy (CCR) and France.

Electronic data processing systems for the supervision of <u>trans-frontier</u>
shipments by rail are at present being developed. In the near future
the national EDP systems for railway transport will be connected to EDP
systems in neighbouring countries (the so-called Project HERMES).

For transport by roads, inland waterways and transport in coastal waters
or on high seas no EDP system, comparable to the SGG system for railroads,
exists in the Netherlands.

A system which can be mentioned for road transport and which has only
recently become operational in the Netherlands is TRADICOM. The system
which was set up by the national Organisation for Road Transport (NOB) is
a databank on transport of dangerous substances connected to the Dutch
VIDITEL-view data system of PTT. TRADICOM can be easily consulted by
every transporter who wants to transport a certain dangerous good, substance or waste. The following data can be obtained:

- UN-number and obligatory requirements for transport
- classification of the substance (class number and figure)
- obligatory routing
- transport prohibitions for certain tunnels
- notification obligations
- labelling prescriptions
- regulations regarding transport by tank-containers
- obligations of orange placard and respective numbers.

The system will in the future also be adapted to international use.
At the moment, approximately 800 firms are connected to the system for use
in commercial road transport.

For maritime transport there is a newly developed data coordinating network (financed by the EEC and the European Association of Port Information)
of European ports like Antwerp, Bremen/Bremerhaven; Genoa; the British Port
Association, Hamburg, Copenhagen, Le Havre and Rotterdam. This system
covers characteristics of ships and ship movements, such as departures,

ship draughts of departure, ship-specific supplements, port of destination
port specific requirements such as trans-shipment, ship averages and accidents. A world-wide expansion would be preferable.

A harmonized EEC system of electronic data processing for trans-frontier
shipment of dangerous wastes can be based on:

- for railway transport: the so-called HERMES project
- for sea transport: the newly developed data coordinating network
  of European ports.

For trans-frontier transport by roads and inland waterways, EDP systems
could probably be set up   on the basis of one of the two above
mentioned systems with the necessary adaptions and improvements. For
transport by road the TRADICOM databank probably can be used in the EDP
system as an example of an information-supplier.

VII; EMERGENCY MEASURES

1. Measures in the event of accidents during the transportation of dangerous wastes (substances) to minimise environmental damage

In case of an accident the first emergency organisation informed is the fire service, the police or an alarm centre.
These services may warn other emergency services and may coordinate the various services.

Usually the fire service is the first service to take measures. These first emergency actions can be based on the knowledge of the driver or on the safety labels at the packing or vehicle. However, the main source for identification of the required emergency actions is the number of safety instructions (e.g. Tremcards) accompanying the cargo (given by the consigner) e.g. fire extinguishing instructions, instructions about how to deal with the spilled wastes, instructions about the correct protective clothing.
The fire service will try to identify the spilled dangerous substance with the help of the UN substance number. At the same time an emergency action code can inform the emergency services about the required emergency actions to be taken (in case the safety instructions, which have to accompany the cargo, are not available). The UN substance number as well as the emergency action code are indicated at the outside of the vehicle.

If required an "emergency centre" may give information (about the identification, the nature of the danger and possible actions) with the help of the UN number.

If the required information is not available via these methods, the producer, disposer or local chemical firm can be asked for information. As the last means for information, a qualified analyst may be used.

This intervention procedure is almost the same for all Member States. However, there are many differences with respect to the emergency organisations, the emergency consultation centres, the emergency action codes and the accompanying documents with safety cards.

2. Emergency and consultation centres

Every country has its own centres, some better developed than others. According to information reveived, the British information system "Chemdata" seems to be an attractive system. It is a databank (developed by NCEC, Harwell) which provides information about 30.000 chemical products. The system consists of a network of microcomputers with a hard disc that is regularly being updated by Harwell.

Almost every fire service has a microcomputer. Available information consists of details about the properties, warnings, emergency actions and a list of chemical companies that can be asked for advice in case of emergency. However, problems will arise with mixtures of chemicals, of which wastes are often composed. For those questions for which Chemdata has no answer, one can try a "longtrip" procedure via a NCEC official in Harwell: this official will try to answer the question with the help of the Harwell databank.

Two other interesting British systems are "Cirus" (which resembles Chemdata) and "CEAS". The latter has been set up by chemical and transport companies. Every participating company provides details about their products to the CEAS databank. A 24 hour CEAS telephone number can be indicated to the vehicles or packings. In case of an accident, one can ask for advice and information (comparable with Chemdata's information) via this telephone number.

In Germany, the police or fire service can get advice and assistance from the emergency service system TUIS: hundreds of chemical facilities are situated throughout the country, which are willing and able to give assistance in the immediate vicinity in the case of an accident.

In the Netherlands, an Alarm Centre for Protection against Accidents with Dangerous Goods, located in The Hague, can give information about dangerous goods, their potential hazards and how to handle them. The Alarm Centre State Police Force in Driebergen can also be consulted. Information is given about the services which can be called for assistance or advice.

The Rotterdam Harbour area has its own special service for advice, information and assistance, the so-called Dienst Centraal Milieubeheer Rijnmond (Centralized Environmental Protection Agency Rijnmond). Furthermore, there is a Dutch Chlorine-Calamities Service.

In Italy the emergency system SIET resembles the British CEAS.

In Ireland one can make use of the British Chemdata or of a similar Irish system (IIRS).

In France, CODISEC acts as a source of information for emergency teams.

The main problem of the most (if not all) emergency consultation centres is their lack of information about mixtures of chemicals and therefore about wastes. It is obvious that a communal effort to obtain (emergency) information about dangerous wastes is desirable.

3. Technical emergency organisations

Usually the main emergency organisation is the fire service and sometimes the civil defence. Furthermore, in the UK there is the "Road Haulage Association" which has set up a scheme (together with the Chemical Industries Association) in which RHA-members will provide an appropriate empty tanker and personnel for the transfer of a bulk load in case of an accident.

4. National alarm and disaster plans

In France, when the dimensions of an accident exceed the resources of local services, the national commissioner for the Department concerned may take the decision to launch the ORSEC plan. This plan has the power to requisition all resources in the Department. This plan includes many annexes, corresponding to the types of accident to be dealt with (p. 115, French national report).

In the Netherlands, according to the Disaster Act, every municipality must make a disaster plan, in which the combating procedure of an accident is established. In such a plan is laid down, for instance, which organisations have to act and their responsibilities.

The municipalities are also encouraged to make up a disaster prevention plan. In such a plan it is evaluated which disasters or accidents are likely to occur in the regarded municipalities, and what should be done in case such an accident occurs. A national disaster plan does not exist.

As a large transport company, the Dutch Railroad Company has its own disaster plan.

5. Transport Emergency Cards

According to the international regulations (ADR), a driver of a road vehicle carrying a dangerous substance must have with him, throughout the journey, clear information in writing on the immediate action that should be taken either by the driver himself or by any other person present in the event of an emergency during transport. Additionally, it is envisaged that the same information in writing should be available to the emergency services (police, fire brigade, ambulance).

The consignor needs to provide the safety instructions to the transporter not later than when giving the order for transportation. This information may be by means of a letter, a telex or transport emergency cards, such as those published by CEFIC (Conseil Européen des Fédérations de l'Industrie Chimique).

These cards (Tremcards) meet the ADR-requirements and bring additional benefits:

- they coordinate the efforts of the chemical manufacturers in European countries to work out appropriate instructions, resulting in a uniform system on a wide range of knowledge and experience in handling the products involved;

- the text of cards of the various dangerous substances are available in the requisite languages for use by manufacturers, senders or carriers;

- the cards are presented in a standard style, phrasing and format, designed to help those who are likely to be first on the scene of a road incident, e.g. driver, police, fire brigade, ambulance, to determine quickly what measures should be taken;

- due regard is also paid to <u>environmental hazards</u> that might arise from accidents or emergencies during road transport, e.g. when the <u>authorities</u> should be <u>warned</u> about possible <u>pollution of water courses,</u> <u>sewers</u> or possible <u>damage to soil and vegetation.</u>

<u>Disadvantages of Tremcards</u>: whilst Tremcards are obviously very useful in dealing with single substance emergencies it should be noted that it is unlikely that a waste would fall into this category, as even slight contamination is likely to affect the degree (if not the category) of risk presented. The <u>unsuitability of the nomenclature for dealing with waste products,</u> necessarily means that <u>the safety cards like the Tremcards are also inadequate.</u> In order to improve this situation, <u>the following options may be envisaged:</u>

a) Creation of specific safety cards:
   This opinion is to be considered in parallel with the introduction of waste products into the nomenclature. However, the variety of the chemical composition of waste products transported means that a precise and comprehensive catalogue might be very unwieldy to use.

b) Obligation on producers to give transporters safety instructions:
   These cards give details on the special precautions to be taken in case of accidents, where these precautions differ, from the safety card.
   A similar duty is already imposed on foreign producers in the form of the declaration of importation of dangerous waste.

The latter procedure, based on a prototype document, would allow people involved to have a safety card suited to each type of waste. It could also serve as a basis for a feasibility study on the introduction of the specific safety cards suggested above.

Tremcards are in use in F, FRG and UK. In France these cards are only being used for the transport of unpackaged dangerous substances. In Ireland, sometimes Tremcards are being used as an alternative for their own kind of safety cards. Denmark has designed and developed its own chemical waste cards. These cards are also useful for wastes, unlike the Tremcards.

6. Emergency action code

An emergency action code may consist of a code number indicating the hazard(s) designated to a substance or material. It may be used for marking dangerous cargos allowing the emergency services to identify the actions required.

All ADR-countries in the EEC, except UK and IRL, use the Kemler system: it consists of two or three digits of which the first indicates the main hazard of the substance. The second and the third provide subsidiary information regarding the hazard of the material and combinations of digits may have specific meanings (Annex VII).

In the UK and IRL chemical transports are marked with the so-called Hazchem code, which is essentially an action code.

The basic difference between the Hazchem code and the Kemler code is that the Kemler indicates the hazard presented by the substance, whilst the Hazchem code indicates the action to be taken in the event of an emergency. Clearly neither code is a substitute for the wealth of information available by computer from the UN numbers (for pure chemicals), but for wastes the Kemler code in many cases merely duplicates the danger symbol on the placard, where Hazchem indicates the type of fire-fighting medium to be used, and also the degree of personal protection required by the fire-fighting crew. Hazchem also indicates the need for containment or dilution of a spill, and whether or not the area needs to be evacuated. It is a problem to fit wastes into categories required by the Hazchem code (consists of only one code). Therefore, there has been at least one code proposed to apply specifically to wastes: Discode (for Disposable Code).

The placard used with Discode incorporates four coloured squares, red, yellow, blue and pink, representing respectively fire, reactivity, toxicity and handling. Numbers of the squares from 0 to 4 for fire and 0 to 3 for the others indicate the degree of hazard presented. For fire the numbers 1 to 4 have the same meaning as the Hazchem numbers, whilst 0 means that the substance will not burn. A more detailed Discode card is expected to be carried in the cab of the transport. It must be remembered that it is, at the moment, only a suggested code and has not yet been put into practice. Given the versality of Hazchem it may well be that Discode will never achieve popularity.

The Hazchem code (in UK and IRL) or the Kemler code (in other countries) are always indicated together with the UN substance number on an orange plate at the front and rear end of a vehicle carrying dangerous substances.

In Denmark, the UN substance number is substituted by the Danish word for waste ("affald") in case the transported dangerous material is not a pure substance but a mixture like a waste. France has a similar marking method.

Probably the easiest way to harmonize the use of an emergency action code inside the EEC is to start from the Kemler system because this system is more internationally accepted than the Hazchem code. However, the Discode system may be interesting because it is specifically designed for wastes.

7. Driver training and licensing

The international regulations (ADR, RID, ADNR, IMDG-Code) make a number of special demands with respect to the training and licensing of drivers carrying dangerous substances, e.g. instruction on preventive measures during transit and emergency actions to be taken by the vehicle personnel immediately following an accident pending the arrival of the competent emergency services.

In general, the regulations of classification and packaging of hazardous wastes within the scope of ADR/RID seem to be rational and practical, but not without a few problems.
One of these is regarding the drivers training in relation to the required basic knowledge on the properties and dangers of hazardous waste.

It is noticed that mixtures or solutions of dangerous wastes can contain small quantities of substances which under normal transport conditions have a minor degree of danger. In case of an accident, however, they may be the source of unexpected development of serious dangers for the vicinity.

A consideration of this statement is outlined in annex VI of this report. On the basis of this consideration a revision of drivers training and licensing requirements is proposed.

In most countries there are some additional regulations in the international agreements.

In Belgium, drivers of road vehicles carrying dangerous substances in tankers or tank-containers are required to obtain a special certificate. The certificate may be obtained by showing practical experience or by attending training courses and after passing a test. For different groups of dangerous substances, different certificates are required.

In the FRG, since 1979 there is an obligatory training and licensing system for drivers of motor vehicles with dangerous goods. According to the tasks several basic courses of instructions are prescribed:

- duty and responsibility of driver, operator, carrier etc.,
- general dangerous properties of substances such as fire, explosion, poison, contamination with water, etc.,
- placarding of hazards and information such as documentation, tremcards, licensing, labelling, orange placard with identifying hazard numbers etc.,
- vehicle equipment including electrical components, fire prevention, fire fighting, flooding of tanks, loading, trailers etc.,
- accident prevention including road wetness, bend driving, braking distances, personal protective devices, first aid etc.

In the NL large industries and transport companies may train their personnel. Drivers of road vehicles get a broader training than required in the ADR; the driver of a transport unit for explosives for which a transport license is required must be sufficiently expert with regard to the nature of the substances, the hazards connected with the transport, the legal regulations and the measures to be taken in the event of delay, accident or other irregularity.

In the UK it is anticipated that when the "Dangerous Substance Regulation 1981" comes fully into force no one will be allowed to drive any vehicle being used for the transport of dangerous substances unless he has received an approved course of training.

With regard to EEC harmonizations, it can be mentioned that the ADR requirements for driver training and licensing will become more stringent and that for this the German training system will be followed. This system can probably also serve as an example for the training systems of other international regulations (ADNR, IMDG-Code, RID).

8. International conventions

Following the task of international harmonization the CDG (Carriage of
Dangerous Goods)-Sub-Committee of the IMO introduced 93 group cards similar to the Tremcards of ADR, entitled "Emergency Procedures for ships
carrying dangerous goods" (adopted by the Maritime Safety Committee in
1980).

They prescribe procedures to be followed in case of accidents involving
dangerous substances. The sub-title is "Group Emergency Schedule" (EmS).
These cards cover all substances and articles of the Classes 1-8.

The recommended emergency procedures and actions refer to substances and
articles listed in the IMDG-Code; they should be followed in conjunction
with the information provided in the Code and the "Medical First Aid
Guide for Use in Accidents involving Dangerous Goods" (MFAG).

The "Group Emergency Schedules" are divided into 5 sections, namely:

section 1 - Group title with the emergency schedule number (EmS No)
section 2 - Special emergency equipment to be carried
section 3 - Emergency procedures
section 4 - Emergency actions
section 5 - First Aid treatment.

The MFAG (adopted 1982 by the Maritime Safety Committee) is based on a
team work of special working groups of IMO, WHO (World Health Organisation) and ILO (International Labour Organisation). An illness following
an accident involving dangerous goods on a ship could be difficult to
diagnose and treat. The "Introduction" of MFAG gives useful information on the appropriate approach.

VIII. EVALUATION OF RESPECTIVE NATIONAL SYSTEMS, INCLUDING POSSIBLE SHORTCOMINGS AND SUGGESTED ALTERNATIVE SOLUTIONS

1. Summary of shortcomings per country

Belgium:

Most criteria being applied to the classification of dangerous commercial goods may be applicable to wastes as well. Since wastes often have specific properties, e.g. due to their heterogeneous nature, certain criteria have to be notified, especially regarding the testing methods to be used. Research may be required in order to estimate certain hazard characteristics of waste.

France:

Waste products present certain characteristics which make the regulations on the transport of dangerous substances inappropriate to these products.

The revision of the RTMD, introducing classification criteria and generic "not specified elsewhere" headings for categories 3, 6.1 and 8 in the nomenclature, makes it easier to classify wastes. However, although assimilation will be simpler, the identification number will correspond only in a distant manner to the waste product. This could lead to mistakes in selecting the assimilation substance, confusion as to the nature of the product transported, and inappropriate measures taken in case of accident. Moreover, the absence of waste products in the nomenclature of dangerous substances hinders the use of transport ministry statistics in the area of the environment.

The unsuitability of the nomenclature for dealing with waste products necessarily means that the safety cards are also inadequate.

Ireland:

The national regulatory system regarding the transport of dangerous wastes is inadequate (there has been no ratification of ADR). The basic problem is that the transport of wastes is not adequately controlled by the national regulations of dangerous wastes, and the national Road Transport regulations do not, in practice, apply to dangerous wastes.

The current dichotomy between the Road Transport regulations and Dangerous Waste regulations leaves a serious gap in the national system of control of dangerous wastes.

The national system of regulation with regard to rail (and inland waterway.) transport of hazardous wastes suffers from similar deficiencies as the road transport regulation system. The basic problem again is the very restricted list of substances covered by rail regulations. Moreover, the regulations in the case of rail transport are still only in draft form.

There has been an inordinate delay too in the drafting of loading and unloading regulations for road transport of dangerous substances.

The Netherlands:
The environmental legislation contains no conditions and regulations with respect to transport of dangerous waste.

It will be clear that in the NL the definition of "dangerous" according to environmental legislation and according to transport legislation differ. There is no coordination with respect to notification on a governmental level between the ministries competent for the enforcement of the environmental legislation and of the transport legislation.

The regulations of the IMDG-Code and those of RID/ADR differ slightly. In practice this may cause difficulties with regard to packings and in an extreme case it could be necessary to re-pack a substance. Furthermore, it is noticed that many substances which are classified into class 9 (N.O.S) of IMDG should be classified in other classes under RID/ADR regulations. Lack of information, however, hinders this reclassification, so that re-collecting of information by telexes to sender, etc., is necessary.

Since loading and unloading is considered in the Netherlands as being an essential part of transport, more regulations with regard to this subject should be adopted in the transport legislation. A possible gap in control activities with respect to transportation of wastes may occur in case an entrepreneur does not comply with the transport regulations (by mistake or on purpose). In most cases such a transport will probably pass unnoticed.

Denmark:

The Danish Local Authorities' Central Treatment Facility for Hazardous Waste (Kom-munekemi) is the relevant authority as regards the treatment and disposal of chemi-cal and dangerous waste in Denmark, but local authorities also have specific regu-lations concerning their own areas. Beyond this, there are many organisations in-volved in various areas pertaining to the conveyance of dangerous wastes, and a certain professional competition exists among the parties involved.

With respect to trans-frontier transport it is necessary for some firms to seek professional advice in order to ensure that all legal requirements are met. In some cases this advice must be paid for, which results in companies being reluctant to seek advice; therefore, Denmark should consi-der the establishment of centralized information centres for trans-frontier transports of dangerous waste.

The FRG:

The waste legislation of the FRG is quite satisfactory. Responsibility for the implementation and the monitoring is delegated to the 10 Länder. Within the framework of federal legislation, the Länder are fairly autono-mous in the regulations they establish on the management of chemical waste disposal. Last but not least, the legal framework guarantees control and monitoring. Federal legislation provides, especially for compulsory noti-fication and record-keeping, a permit system and a consignment note-system (Begleitschein). The utilization of wastes includes the whole pro-cedure of collection, transport, treatment, storage and tipping.

The primary condition for waste disposal is that it must be transported as a hazardous substance. The new requirements in the harmonized RID/ADR re-gulations for land transport of dangerous wastes in tested and periodically inspected packaging including their definition and classification creates the basis for this condition and for public security during their carriage. Furthermore, driver training for transport of dangerous goods in tanks and tank-containers became obligatory with great success in 1977.

Italy:

There is conformity between the international transport agreements and the Italian legislation (e.g. consignment note), but there are particular developments concerning the prevention of injury, the protection of water

from pollution or legal domestic measures classifying and regulating
the packaging and labelling of dangerous substances and preparations,
which conflict at present with the definitions of toxic and dangerous
waste. In addition, the regional authorities are entitled to issue regulations on the subject, although these may not conflict with legislation
enacted at central and national level.

In the specific case on the carriage of dangerous goods, Italy has set up
an ad-hoc research group, by agreement with the competent UN agencies
(e.g. ECE-UN Transport Division),   to reconsider the problems of
packaging in relation to the risk to human life and limb; the aim is to go
beyond the present phase in which, when defining toxicity, reference is
made to the risk to animals. Furthermore, in case of dangerous waste
Italy is engaged with intoxication problems as mentioned under health
hazards (acute and chronic), with risks for the environment and human life.
These problems and also "Threshold Limit Values", which concern only workers in facilities, are not considered within this area.

There are great differences in collection, disposal or conveyance of
toxic and dangerous wastes. Their transportation must be seen only in
connection with being safely packed (according to the relevant international and national regulations) and kept in tested, approved and periodically inspected packaging (e.g. drums, tanks, tank-containers), which are
sealed and tight under normal transport conditions. In this case an
intoxication of persons is unrealistic.

United_Kingdom:
The principle piece of legislation pertaining to waste disposal is the
"Control of Pollution Act 1974" and for hazardous wastes the Control of
Pollution (Special Wastes) Regulations 1980" (the list of the EEC Directive
319/78 was added). As in other countries, under these "Special Waste Regulations" the greatest safeguard for avoiding violation of dangerous waste laws
is the documentation procedure. Fortunately, the paperwork involved in
these regulations ensures that if a load is dispatched and does not arrive,
the disposal authority is aware of this and can take steps to locate the
waste.

Furthermore, chronic effects which are legislated for in the "Classification, Packaging and Labelling of Dangerous Substances Regulations 1984"
or genetic effects in connection with toxic wastes (especially in landfill
sites) could be better controlled by a small number of regional authorities
concerned with hazardous waste management in geographically determined areas.

The ADR agreement has been ratified by the UK, therefore international
trans-frontier shipments by road in this country are permitted. For
their domestic traffic there are three regulations concerning classification and labelling of dangerous substances, their conveyance by road
in road tankers and tank-containers and the conveyance by road in packages.

These regulations are very effective in that should the contents escape,
an offence will clearly have been committed. With regard to British law,
the strict liability under these regulations means that the best a lawyer
can hope for, following a spillage which is not the result of some
external force, is to minimize the penalty to his client rather than secure his acquittal. When the draft regulations relating to dangerous
goods in packages etc. become law, it will effectively mean that all road
transport of dangerous goods must be conducted in a safe manner, with
appropriate penalties for transgressions.

At present, the three above mentioned regulations are not quite comparable with the ADR and RID agreements regarding packaging requirements
(performance testing, approval, periodical inspections etc.), safety
equipment (fire extinguisher etc.), placarding,driver training (ADR),
loading and unloading operations etc. Nevertheless, within a transitional period of 5 years the UK will fall into line with the ADR/RID
regulations.

2. Review and conclusions

2.1. Classification

Most EEC Member States have given a definition or a circumscription of
dangerous waste in their respective environmental legislation. Subjective and objective criteria are then combined. The boundary between
"waste" and "product" or "good" is vague and fluctuates.

Various interpretations are possible, depending on e.g. market situation, commercial value and possibilities for recycling. There is no uniformity in approach and classification of dangerous waste in relation to environmental protection in the EEC Member States. Differences in approach and classification are marked by use of different characteristics, differences in consideration and assessment of dangerous properties and differences in risk assessment.

Some classification systems are of the qualitative type, others are quantitative. Waste lists are also applied.

It is therefore not surprising that the boundary between normal waste and toxic or dangerous waste differs from country to country.

The various national transport legislations show a more consistent approach in classification of dangerous substances since they are largely based on international agreements.

With respect to the identification of the meaning of degree of danger according to transport legislation, it is absolutely irrelevant whether goods, substances, solutions or mixtures are being considered as waste or not.
Hence, until very recently, no separate classification system for the identification of the particular classes of risk for waste existed, in for instance RID/ADR.

Recently the RID/ADR classification system became more harmonized with the UN-classification method. It is also possible now to classify solutions and mixtures (including wastes).

Furthermore, a ninth class for environmentally dangerous substances (PCB, asbestos) is envisaged in the RID/ADR agreements.

It must be emphasized that transport regulations only consider immediate exposure to dangerous substances, while environmental approaches primarily take into account the after-life of products in the environment, and the harmful effects that can derive from long term exposure.

It is likely, that this disharmony between transport legislation and environmental legislation on the one side and differences in approach among the various environmental (national) laws on the other, will lead to all kinds of problems, misunderstandings and conflicts of interests.

Therefore, we believe that "waste tourism" and consequently impacts on health and environment cannot be excluded under the existing laws and regulations.

This statemant is sub-structured by the following remarks:

1) In conformity with Directive 84/631/EEC trans-frontier movements of hazardous waste should be executed by the application of international transport regulations (art. 8). The Directive, however, only applies to PCBs as described in art. la of Directive 76/403/EEC and to toxic and dangerous wastes containing one or more of the substances listed in the appendix to Directive 78/319/EEC. It excludes chlorinated and organic solvents (art. 2a).

   Wastes which are destined to be re-used, regenerated or recycled are also exempted (art. 17).

   It will be obvious that the trans-frontier movements of several dangerous and/or toxic wastes are not covered by Directive 84/631/EEC.

2) Since minimum concentration levels for dangerous wastes differ among countries (e.g. Belgium and the Netherlands) it is possible that a waste being defined as dangerous in one country, is considered as harmless in another country.

   In the latter case the exact concentration levels in "dangerous wastes" are not always known since they are not required under environmental law.

3) It may be assumed, that a comprehensive analysis of the composition of waste is difficult to obtain, even in case concentrations and levels are required for classification purposes under environmental law.

Recent Dutch investigations have demonstrated that it is difficult or even impossible to set a mean and standard deviation for concentration levels of dangerous substances in wastes.

Experiences with the Dutch notification system have also shown that analysis of wastes in practice are focussed on the requirements of waste converters.

For instance, in the case of incineration the predominant criteria for acceptance are caloric value, presence of halogens (protection of oven) and /or certain heavy metals (emission limits).
It will be obvious that the generator tends to limit the amount of expensive analysis for waste, which is worthless or has a negative value.

Sometimes analyses costs are avoided by a simple designation as "dangerous waste". The obligatory notification is then fulfilled by reporting an estimation of the presence of certain substances above the set minimum concentration level.

There is no reason to assume that waste will be "analysed" in a different way under transport regulations.

4) Even if exact concentration levels are known, cases may be assumed that a transport of hazardous waste is not ruled by transport regulations.

For instance, a waste containing 60 ppm mercurisulfide is considered as a dangerous waste under Dutch environmental law (minimum level 50 ppm).

Conform ADR margin number 2601, sub 53 this substance is exempted.

5) For waste generators, there are several possible ways of deciding upon classification. For example, in the case of toxic and harmful wastes approaches may include:

- either the LD_{50} or LC_{50} of the waste determined by tests;
- or the calculation of the LD_{50} or LC_{50} of the waste by using the following formula:

$$LD_{50} \text{ or } LC_{50} = \frac{100}{\sum_{i} \frac{Mi}{Li}}$$

where Mi is the percentage by weight of ingredient i in the waste, and Li is the LD_{50} (oral or percutaneous) or LC_{50} (via the respiratory tract) of ingredient i in the waste.

Where a waste can have either a toxic or a harmful effect the most penalizing classification arrived at by these calculations must be applied

- or the de facto consideration that the waste is dangerous in case it is not wished to use either of the two methods, set out above (severe approach).

It is clear that the first two classification approaches are expensive and time consuming.

Standardisation using hypothetical LD_{50} or LC_{50} values may be difficult because of strong fluctuations in composition.

A disadvantage of the last approach may be the costs for elimination of the waste, although it may be assumed that the transport costs of dangerous waste and normal waste are comparable.

2.2. Packaging

The rearrangements of the recent RID/ADR regulations on the basis of the
UN-Recommendations "Transport of Dangerous Goods" will be of great importance for world-wide trade. This harmonization development will result
in uniform requirements used for all modes of transport. With the integration of the chapter 9 (General Recommendations on Packaging) of the
UN-Recommendations into the relevant Annex V of RID and Annex B.5 of ADR,
it is now possible to transport in border traffic dangerous goods or
wastes without difficulties. The advantages are:

- Simplification of Hazardous Materials Regulations for shipments;
- a significant reduction in the volume of the regulations;
- provisions for greater flexibility in the requirements for the design
  and construction of hazard materials packaging in order to recognize
  technological advancements in packaging;
- the promotion of safety in transport through the use of better packaging;
- a reduction for the need of exemptions; and
- the facilitation of international commerce.

1) These packaging requirements consist of specifications, performance
   tests, marking etc. In the case of new manufactured receptacles the
   FRG, France, Italy, the Netherlands and in several details Belgium
   are in accordance with these requirements. The UK needs for transition
   until 1990. The required tests in the UK will be carried out in accordance with the relevant parts of ISO Standard 2248, respectively British Standards 4826 (BS). ADR has not been ratified by Ireland; although for 25 substances they have specific testing requirements of
   receptacles.

2) A new chapter 16, titled "Recommendations on Intermediate Bulk Containers" (IBCs) will be added to the UN-Recommendations in 1987. These
   requirements are going to form a new Annex in e.g. ADR and RID and will
   give the user the possibility for safe conveyance of dangerous goods
   and wastes. Some EEC Member States - the UK and the FRG - are already
   using these container types with success. Furthermore, there is a

development, e.g. France, the UK, the Netherlands and the FRG of a
new IBC-type produced by rotational moulding of powdered thermoplastic
material.

3) For economical transport of multitudes of receptacles the freight
container type is suitable and also of importance for the safe conveyance of dangerous goods and waste. They must be designed, constructed, tested and approved under the ISO Standard 1496 (in the UK: BS
3951; in France: UIC 590) and the "International Convention for Safe
Containers" (CSC). In Belgium, Denmark, the FRG, France, Italy, the
Netherlands and the UK these requirements apply.

4) Due to their design and construction, the tank-containers and tanks
are one of the safest means for transport of dangerous wastes. On the
basis of the UN-Recommendations, chapter 12 (Recommendations on Multimodal Transport), the tank-containers for sea-transport are regulated
in the IMDG-Code as so-called IMO-types I, II, IV and V and for land
transport in Annex B.lb/ADR and Annex X/RID. The relevant requirements
of ADR/RID, IMDG-Code and ADNR apply in Belgium, Denmark, the FRG, Italy
and the UK; in Ireland only in the case of RID and IMO. The tanks
are regulated by Annex B.la/ADR and Annex XI/RID. In Belgium, the
FRG, France, Italy, the Netherlands and the UK these regulations apply.

5) The Consignment Procedures, such as the marking of packaging and
labelling, are obligatory in most of the EEC Member States. In the case
of documentation of wastes there is a difference between the international transport regulations and the relevant EEC-Directives respective
the different country legislations, that is based on the special environmental and waste regulations.

6) In Belgium, Denmark, the FRG, France, Italy and the Netherlands the UN-
Recommendations and the specific RID/ADR requirements provide the
basis for the placarding system. The UK and Ireland have their own
special system -Hazchem system. In the first moments after an
incident this system is the best informative one, but with reference
to the international transport regulations only applicable within their
territories.

7) World-wide there is a lack of <u>international placarding for the transport of toxic and dangerous wastes</u>, especially for the trans-frontier shipment in the EEC. Based on their languages, France uses "déchets", Denmark, the FRG and the Netherlands the letter "A" and the UK the letter "W".

8) <u>Special tankcars and attached tanks</u> are of importance for the carriage of toxic and dangerous wastes and therefore they are used in <u>Belgium</u>, <u>France</u>, the <u>FRG</u>, <u>Denmark</u>, <u>Italy</u> and the <u>UK</u>. A new ADR requirement especially for vacuum-pressure-tanks is in the drafting stage and will be enforced by the end of 1987.

9) In several Classes of the ADR and RID regulations the <u>methods and practices of the cleaning of receptacles</u> are prescribed. Depending on the properties of the dangerous substances or wastes, it shall be done with water and/or steam or special solvents. The requirements in the IMDG-Code are similar. Except for the FRG and the Netherlands there are no specific provisions in national legislation and no reference to international agreements in Belgium, Denmark, France, Ireland, Luxembourg and the UK. Denmark has a Ministerial Order for the Protection of Workers.

2.3. <u>Means of transport</u>

Most toxic and dangerous wastes are moved by road. In this context there are differences in transport for waste production, recycling (in companies facilities) or disposal (landfill site). Rail transport is more specific and depends on the quantity; countries with islands or coast States are also involved in sea-transport; nevertheless, packaging, equipment etc. must be in conformity for all modes of transport.

1) <u>Electronical devices and systems, fire extinguishing equipment, special safety equipment</u> (e.g. warning lamps), the <u>supervision during parking periods, cooling and ventilation equipments</u> are prescribed in the ADR regulations. All EEC Member States as ratificants or ADR (with the exception of Ireland) are liable for compliance with these relevant requirements.

2) Of importance are <u>testing, certification and inspection</u>. The packaging, containers, tank-containers and tanks should be designed, manufactured and tested under a quality assurance programme which satisfies the competent authority, in order to ensure that each manufactured receptacle meets the relevant requirements. After performance testing, approval and certification to the satisfaction of the competent authority every receptacle should be inspected before it is put into service, and thereafter at intervals depending on the specific transport requirements and with regard to

- conformity to design type, including marking;
- internal and external condition; and
- proper functioning of service equipment.

3) Experts or examiners of State administration or from the competent authorized national or international (e.g. classification agencies) institutions are responsible for testing and inspection. Nevertheless, <u>EEC Member States are not at all in line with these international requirements concerning testing, approval and certification</u>, and also the <u>intervals of the inspections</u>. In relation to the international agreements the differentiations are as follows:

- <u>receptacles</u>: Here is a certain conformity in Belgium, Denmark, the FRG, France, Italy the Netherlands and the UK.

- <u>road vehicles</u>: Belgium, France, the FRG, Italy and the Netherlands are in conformity, whereas in Denmark the traffic police are responsible; in Ireland certain garages are authorized, and in the UK authorized garages may carry out roadworthiness tests on vehicles whose gross plated weight does not exceed 3.5 tonnes (including private cars) but all heavy goods vehicles must be tested annually at a Department of Transport testing station for roadworthiness.

- <u>railway</u>: Experts of the railway administrations have to carry out these examinations in Denmark, the FRG, France, Italy, Ireland, the Netherlands and the UK.

- <u>inland waterway vessels</u>: There is a certain conformity within the FRG, France and the Netherlands.

- <u>sea-going vessels</u>: Belgium, Denmark, the FRG, France, Ireland, the Netherlands and the UK conform with eachother.

4) The land transport of dangerous wastes is regulated by the recent edition of the RID/ADR regulations (1985) and by the relevant EEC-Directives and the specific domestic environmental legislation. The <u>water-borne transport of dangerous wastes</u> differs within the EEC Member States. Most of the coastal countries use the sea-transport for dumping operations or for incineration purposes. According to the "Convention on the Prevention of Marine Pollution by Dumping of Wastes and other Matters" (MARPOL 73), these actions will continue to decline. The FRG will stop these operations in the nineteeneighties. In all States these dumping operations are controlled and permitted by the relevant competent authority. Another kind of sea-transport of wastes is found from Corsica to the continental France for treatment purposes.

5) The <u>combined traffic</u> in land transport is a so-called pick-a-back operation of e.g. a tankcar on a railway lorry (in USA:"Piggy-back"; in Western Europe: "Rail-Route", "Semi-Remorque", etc.). House-to-house traffic would be easier and rationalization could be the result. Design and construction of these vehicles are being constantly improved. They are more or less unused in the transporting of toxic and dangerous wastes, nevertheless, a trans-frontier shipment at sea for waste treatment and disposal purposes by a combined traffic, sometimes including the "Piggy-back"-system, could be important. The so-called "short international voyages"" in this case are interesting and present an appropriate method. The "Memorandum of Understanding by short voyages", agreed between Denmark, Sweden, Finland and the FRG, enforced in 1980, is typical for this method of sea-transport. The conveyance of dangerous goods or wastes on sea-going vessels will be carried out under the relevant requirements of land transport (RID/ADR) and not by the more restrictive one of sea-transport (IMDG-Code), but only under the conditions of a short voyage in the Western Baltic Sea.

The IMO has been consulted on the problems regarding the "short sea voyages" world-wide, and as a result, new developed requirements are in draft stage of preparation. The European arrangements for "short international voyages" e.g. in the British Channel, between the UK and Ireland or in the Mediterranean Sea between Corsica, Sardinia, Majorca and the Continent, will be of interest.

2.4. Loading and unloading

In the ADR and RID agreements   loading and unloading operations are regulated; especially the filling rates of tanks and tank-containers. If
dangerous goods or wastes are involved in loading and unloading procedures, there are some basic requirements for security which would apply
to most of the different classes of the ADR, for instance;

- the attendance of a qualified person during these operations;
- definition of "qualified";
- the quality of a delivery hose or flexible pipe, its use, tests
  and control measures;
- the prohibition of loading and unloading operations in public
  places and areas;
- the prohibition of smoking, fire and ignition sources during these
  loading and unloading operations;
- the duty to set handbrake during these operations;
- secure packages in vehicles against movement;
- prevent relative motion between containers;
- the prohibition to use inflammable materials for stowage.

In addition to these specific requirements of loading and unloading
operations, there are particular industrial legislations for worker
protection regarding these matters in Denmark, the FRG, Ireland, the
Netherlands and the UK.

2.5. Control and monitoring systems

The control and monitoring systems for transport of dangerous wastes vary
among countries; not every country has implemented a trip-ticket system,
and these systems which are already in use are not identical.

There is a lack of legal regulations obliging the notification of the
country of destination (and transit) in case of export of dangerous wastes.
Furthermore, there are no obligations for the holder of the wastes to make
notifications with respect to insurance and safety.

The Directive 84/631/EEC concerns the implementation of a uniform noti-fication system which obliges the notification of insurance and safety and the notification to the country of destination (and transit).

The main shortcomings of this system are the limitation of the list of dangerous wastes for which it applies and the fact that trans-frontier shipments only are concerned.

2.6. Use of electronic data processing systems for the organisation and supervision of the trans-frontier shipment of dangerous wastes (substances)

Until now, there have been no EDP systems for the supervision of trans-frontier shipments of dangerous wastes. The only recent developments concern rail transport (the HERMES project) and sea-transport (a data coordinating network of some important European ports, which covers characteristics of ships and ship movements).

2.7. Emergency measures

The main problem with respect to emergency consultation centres is again the lack of information about mixtures of chemicals and therefore about wastes.

There is also a lack of adequate international alarm and disaster plans.

The unsuitability of the nomenclature for dealing with waste products ne-cessarily means that the safety cards like the Tremcards are also inadequate for wastes.

Regarding the used emergency action codes, one can say that the more gene-rally accepted Kempler system gives too little direct information about the required emergency actions. As previously stated the additional information available from the Hazchem code may overcome many of these problems.

Finally, sometimes the requirements for driver training and licensing are not stringent enough, especially in case of transport of mixtures or solu-tions of dangerous wastes which contain small quantities of substances which under normal transport conditions have a minor degree of danger; however, in case of an accident, they may be the source of the unexpected development of serious dangers for the area.

3. <u>Recommendations</u>

* Harmonization of environmental and transport legislation, and the legislation of different countries e.g. with respect to definition of dangerous wastes.

* Also with respect to the previous point, a comprehensive and internationally accepted uniform list of dangerous wastes should be developed. This list must be included in (new) class 9 (N.O.S.) of the international transport regulations for transport of hazardous sustances/wastes. The holder of the waste must follow the transport regulations when the waste is mentioned in this list, unless he is able to prove that his waste is not dangerous according the classification criteria for dangerous substances/wastes (e.g. concentrations are below certain minimum limits). This list could be a harmonized version of the lists of dangerous wastes which are already in use in some countries (in their environmental legislation).

* The acceleration of the harmonization developments in relation to the international UN-based transport agreements should be encouraged. There are particularly the following inconsistencies:

  - the construction and use of tested packaging,
  - their approval and periodical inspection,
  - uniform placarding measures for the transport of dangerous wastes,
  - fire and safety equipment,
  - loading and unloading operations.

* Development of a guide (cards system) and/or data base for all dangerous wastes, with information about name and origin, identification number, properties of dangers, classification, required emergency actions, required labelling and addresses to be consulted in case of emergency.

* Development of an emergency action code which contains more direct information about required emergency actions. Moreover, the emergency action code must also be suitable for wastes (e.g. Discode system).

* Developments of Tremcards for dangerous wastes (in combination witn the guide).

* More stringent requirements for driver training and licensing, especially for transport of wastes containing small quantities of substances which are not dangerous under normal transport conditions but which may cause severe dangers in case of an accident.

* Common international effort and consultation for the development of supervising of EDP systems for trans-frontier shipment of dangerous substances/wastes.

* Development of adequate national disaster and alarms plans.

ANNEXES

ANNEX I

Outline of relevant international regulations and conventions on dangerous transports

UN-Recommendations
The United Nations Committee of Experts on the Transport of Dangerous Goods (hereinafter referred to as the "UN Committee") was established by a resolution of the United Nations Economic and Social Council on April 15, 1953, (ECOSOC Resolution 468 G (XV)) and first met in 1954.

The United Nations Committee of Experts on the Transport of Dangerous Goods is responsible for the development of recommendations dealing with the multimodal transport of dangerous goods, including provisions for the classification, labelling and packaging of these materials. The Committee, which reports directly to the UN Economic and Social Council (ECOSOC), is currently composed of members from Canada, France, Federal Republic of Germany, Italy, Japan, Norway, Poland, the United Kingdom, the United States of America and the Soviet Union. The Recommendations of the Committee are published by the UN in a volume entitled Transport of Dangerous Goods.
ECOSOC, by resolution, has urged all member states, regional economic commissions and international organizations to bring their practices for the tranport of dangerous goods into conformity with the UN Recommendations.

At the time of the establishment of the UN Committee there were three basic codes governing the international carriage of dangerous goods. The first was the International Regulations Concerning the Carriage of Dangerous Goods by Rail (RID) which are included in Annex 1 of the International Convention Concerning the Carriage of Goods by Rail (CIM), established in 1890. The second was the regulations of the Interstate Commerce Commission (ICC) in the United States which were first issued in 1908 and were used by countries in North America and largely reproduced by the International Air Transport Association (IATA) in its restricted articles regulations. The third was the British "Blue Book" which was broadly recognized as the basic standard for the international transport of dangerous goods by sea. Each of these codes was based on a different approach relative to classification and labelling of dangerous goods.

On 30 September 1957, before recognition of the UN Recommendation by
any official body, the European Agreement Concerning the International
Carriage of Dangerous Goods by Road (ADR) was concluded under the
auspices of the Economic Commission for Europe (EEC). These regulations
largely reproduced the provisions of the RID regulations in order to
facilitate intermodal transport within Europe. This action is
understandable in light of the strong desire for harmony between the
rail and highway modes within Europe and because the UN Recommendations
were not yet fully developed.

In 1965, a significant action was taken by the International Maritime
Consultative Assembly in Paris when it adopted the International Maritime Dangerous Goods Code. The code was based primarily upon recommendations of the UN Committee. Since adherence to international maritime
rules is necessary for any nation involved in maritime trade, the IMCO
action stimulated increased interest in the work of the UN Committee
and its efforts of dangerous goods.

More recently, the International Civil Aviation Organization's (ICAO)
Dangerous Good Panel has developed a complete set of regulatory standards for air transport of dangerous goods. These standards are based
almost entirely upon recommendations of the UN Committee and became
effective on 1 January 1984.

European Agreement Concerning the International Carriage of Dangerous
Goods by Road (ADR)
The European Agreement Concerning the International Carriage of
Dangerous Goods by Road (ADR) was first published in 1959. It consists
basically of three sections. The first section contains the actual
Agreement and protocol of signature. The second section, Annex A to the
Agreement, contains the provisions   concerning dangerous substances
and articles; that is, the list of substances allowed to be carried as
well as the requirements for labelling and packaging of the materials.
The third section, Annex B to the Agreement, sets forth the operational
requirements for carriage by road as well as the requirements for
transport vehicles, cargo tanks and portable tanks transporting
dangerous goods.

The ADR is administered through the Inland Transport Committee of the
United Nations Economic Commission for Europe. Over the past several
years, the primary thrust of amendments to the ADR has been to incorporate the Recommendations of the UN Committee, and amendments incorporating substantial portions of the UN Recommendations will become
effective on 1 January 1985.

Based on the regulations in ADR there is a similar draft-convention for transport on inland waterways; the European Agreement concerning the International Carriage of Dangerous Goods by Inland Waterways (ADN). The ADN, however, never came into force. In 1970 a separate agreement has concluded for the transport of dangerous goods on the Rhine (ADNR), for which the ADN served as a model.

International Regulations Concerning the Carriage of Dangerous Goods by Rail (RID)

The International Regulations Concerning the Carriage of Dangerous Goods by Rail (RID) are found at Annex 1 to the International Convention Concerning the Carriage of Goods by Rail (CIM). CIM is an international convention; however, it is not of worldwide applicability since its provisions apply to only the European Nations which are signatories. The convention is administered by the Central Office of International Rail Transport (OCTI) in Berne, Switzerland. OCTI is not associated with the United Nations. However, RID and ADR conduct joint meetings frequently in an ongoing attempt to maintain harmony between the conventions governing road and rail transport in Europe. The majority of the work of the Joint meetings in recent years has been directed toward the introduction of the UN Recommendations into both the RID and ADR. Consequently, as is the case with the ADR, amendments to the RID incorporating substantial portions of the UN Recommendations will become effective in the nearest future.

International Maritime Dangerous Goods Code (IMDG)

The IMO Subcommittee on the Carriage of Dangerous Goods (CDG) is one of a number of subcommittees of the Intergovernmental Maritime Consultative Organization's Maritime Safety Committee. The CDG Subcommittee is responible for the development of recommendations dealing with the transport of packaged and bulk solid dangerous goods by sea. The Subcommittee has published a comprehensive set of recommendations known as the International Maritime Dangerous Goods (IMDG) Code which deals with all aspects of the transport of packaged dangerous goods by sea (i.e. packaging, marking, labelling, classification, documentation, placarding, stowage, segregration and handling of dangerous goods). The basic systems of listing, classification, labelling and packaging appearing in the IMDG Code are derived from the recommendations on the UN Committee of Experts on the Transport of Dangerous Goods.

The IMDG Code is widely recognized throughout the world as the minimum standard for transport of dangerous goods by sea. It has been fully adopted into the regulations of at least 34 Nations and is in various stages of adoption in many more.

ANNEX II

Classification in accordance with the Recommendations of the Committee of Experts on the Transport of Dangerous Goods.

The UN Recommendations have been or are being embodied in various international regulations on transport with which this report is concerned (RID, ADR, ADN and the IMO Code).
Under the UN Recommendations, dangerous goods are listed in nine classes depending on the type of risk to which they give rise.

CLASS 1:
Class 1 comprises explosive substances, devices containing explosive substances, pyrotechnic articles, etc. As pointed out in the preface, however, this subject does not come within the sphere of our report.

CLASS 2:
For the purpose of transport, a substance is included in this Class if:
a) it has a critical temperature of less than 50°C, or
b) it exerts a vapour pressure of more than 3 kg/cm^2 at 50°C, or
c) it exerts an absolute pressure of more than 2,8 kg/cm^2 at 21,1°C, or
d) it exerts an absolute pressure of more than 7,3 kg/cm^2 at 54,4°C, or
e) it exerts a Reid vapour pressure of more than 2,8 kg/cm^2 at 37,8°C.

CLASS 3:
Class 3 comprises inflammable liquids, mixtures of liquids or liquids containing solids in solution or suspension (although account must be taken of the characteristics which may cause substances to be otherwise classified). Substances come in this Class if they give off inflammable vapours at temperatures of not more than 60,5°C (closed-cup test) or not more than 65,6°C (open-cup test).

CLASS 4:

Class 4 has 3 divisions:

Division 4.1: inflammable solids, other than those classed as explosives (Class 1) which, under conditions encountered in transport, are readily combustible or may cause or contribute to fire through friction;

Division 4.2: substances liable to spontaneous combusion, i.e. those liable to spontaneous heating under normal conditions encountered in transport or to heating in contact with air;

Division 4.3: substances with, in contact with water, emit inflammable gases, i.e. those liable, by interaction with water, to become spontaneously inflammable or to give off inflammable gases in dangerous quantities.

CLASS 5:

Class 5 is made up of two divisions:

Division 5.1: oxidizing substances which, while in themselves not necessarily combustible, may cause or contribute to the combustion of other material, generally by yielding oxygen;

Division 5.2: organic peroxides, i.e. organic substances containing the bivalent -O-O- structure where one or both of the hydrogen atoms have been replaced by organic radicals. They are thermally unstable substances which may undergo exothermic self-accelerating decomposition.

CLASS 6:

Class 6 is made up of two divisions:

Division 6.1: toxic substances, i.e. those liable to cause death or serious injury or to harm human health if swallowed or inhaled or by skin contact;

Division 6.2: infectious substances i.e. those containing viable micro-organisms or their toxins.

CLASS 7:

Class 7 comprises radioactive substances, defined as any substances of which the specific activity is greater than 0,002 microcurie per gram. Class 7, like Class 1, does not come within the purview of this report.

CLASS 8:

Class 8 comprises corrosibles, i.e. those substances which, by
chemical action, will cause severe damage when in contact with
living tissue or, in the case of leakage, will materially damage
or destroy other freight or the means of transport.

CLASS 9:

Class 9 comprises substances which, during transport, present a
danger not covered by the other eight classes.
For example, polychlorinated biphenyls are assigned to Class 9 (UN
No. 2315).

ANNEX III

Modified classification system of ADR/RID

The criteria for classification of substances in the several
classes according to RID/ADR are rigidly changed starting from
1-5-1985 and are harmonized with the recommendations of the UN-
Committee of Experts (orange book). In principle the substances of
the classes 3, 6.1 and 8 are divided in groups corresponding with
the characters a, b and c:
a= very dangerous substances
b= dangerous substances
c= less dangerous substances.

The modified system is based on the following documents of the
joint RID/ADR-Committee;
- OCTI/RID/GT-III/587 (TRANS/GE.15/AC.1/R.255) of November 15, 1984
- OCTI/RID/GT-III/595 (TRANS/GE.15/AC.1/R.263) of November 24, 1984
- OCTI/RID/GT-III/CRP.19/Add.7 (TRANS/GE.15/AC.1/CRP.19) of
  March 29, 1985.

In this respect the following text relates to 2002(8) of ADR/3(3)
of RID:

The following provisions apply to solutions and mixtures (such as
preparations and wastes) not mentioned by name in the lists of
substances of the various classes:

Note: Solutions and mixtures containing one or more components of
      a restrictive class are only permitted for transport, if
      these components are mentioned by name in the list of
      substances of this restrictive class.

(a) Solutions and mixtures containing one component subject to
    RID/ADR shall be regarded as substances of RID/ADR if their
    concentration is such that they continue to present the danger
    inherent in that component itself. Such classification shall
    be carried out according to the danger characteristics of the
    various classes.
(b) Solutions and mixtures containing several components subject
    to RID/ADR shall be placed under an item and letter of the
    appropriate class in accordance with their danger
    characteristics.

Such classification according to the danger characteristics
shall be carried out as follows:

1) Determination of the physical, chemical and physiological
   characteristics by measurement or calculation, followed by
   classification according to the criteria of the various
   classes.

2) If such determination is not possible without giving rise
   to disproportionate cost or effort (as for some kinds of
   wastes), such solutions and mixtures are to be placed in
   the class of the most dangerous component.

Classification according to the most dangerous component shall
take into account the following:

2.1 If one or more components falls within a respective class and
    the solution or mixture presents the danger inherent in that
    component, the mixture or solution shall be placed in that
    class;

2.2 If the mixture or solution contains components falling within
    several restrictive classes and the solution or mixture
    presents the danger inherent in each of those components, the
    mixture or solution shall be placed in the class of the predo-
    minant component; if no component is predominant, the classi-
    fication shall be based on the following order of predominance
    of classes: 1A, 5.2, 2, 4.2, 4.3 and 6.2;

2.3 If components fall within several non-restrictive classes, or
    if in cases of paragraphs 2.1 and 2.2 the solution or mixture
    does not present the danger inherent a restrictive class, the
    solutions and mixtures shall be classified taking into account
    the danger characteristics of such components and their con-
    centrations.

2.3.1 Classification shall take into account the various com-
      ponents and the order of predominance of danger indicated
      in table 1 below. For classes 3, 6.1 and 8, the degree of
      danger presented by the components according to the cri-
      teria of those classes as designated (a), (b) or (c) (see
      marginals 300 (3) RID/2300 (3) ADR, 600 (1)/2600(1), and
      800 (1)/2800 (1) shall be taken into account.

Table   Analytical criteria for waste classification

CLASS of risk	4.1	5.1	6.1 (a)	6.1 (b)	6.1 (c)	8 (a)	8 (b)	8 (c)
3 (a)	sol liq 4.1 3 (a)	3 (a)	3 (a)	3 (a)	3 (a)	3 (a)	3 (a)	3 (a)
3 (b)	sol liq 4.1 3 (b)	3 (b)	3 (a)	3 (b)	3 (b)	3 (a)	3 (b)	3 (b)
3 (c)	sol liq 4.1 3 (c)	3 (c)	6.1 (a)	6.1 (b)	3 (c)	8 (a)	8 (b)	3 (c)
4.1		sol liq 4.1 5.1	6.1 (a)	6.1 (b)	sol liq 4.1 6.1 (c)	8 (a)	8 (b)	sol liq 4.1 8 (c)
5.1			6.1 (a)	6.1 (b)	5.1	8 (a)	8 (b)	5.1
6/1 (a)						6/1 (a)	6/1 (a)	6/1 (a)
6/1 (b)						8 (a)	sol liq 6.1 8 (b) (b)	6/1 (b)
6/1 (c)						8 (a)	8 (b)	8 (c)

Legend:

3.	inflammable liquids	(a)	high danger
4.1	inflammable solids	(b)	medium danger
5.1	oxidizing substances	(c)	low danger
6.1	toxic substances	sol	solids
8.	corrosives	liq	liquids

The footnotes of the table 1 are as follows:

(1) Mixtures and solutions which may have explosive properties (see marginals 3(3) (a) RID/2002 (8) (a) ADR). Such mixtures and solutions shall be accepted for carriage only under the conditions of class 1A.

(2) If the solutions or mixtures contain substances of class 3 marginal 301/(2301), 12° or 13°, they shall be placed in that class under those items.

(3) If the solutions or mixtures contain substances of class 6.1, marginal 601/(2601), 1° to 3°, they shall be placed in that class under those items.

(4) If the solutions or mixtures contain substances of class 8, marginal 801/(2801), 24° or 25°, they shall be placed in that class under those items.

(5) If the solutions or mixtures contain substances or preparations, used as pesticides, of class 6.1, marginal 601/(2601) 71° till 88° they shall be placed in that class under those items, if the determining percentage for classification under letter (c) of the active substance of the pesticide is present.

Example to explain the use of the table:
A mixture consisting of an inflammable liquid classified under class 3, letter (c), a toxic substance classified under class 6.1, letter (b), and a corrosive substance classified under class 8, letter (a).

2.3.2 (RID/ADR) Classification under an item of a specified class in accordance with 2.3.1, shall take into account the danger characteristics of the various components of the solution or mixture. The use of items containing a non-specific collective heading (class 3, 20° and 26°; class 6.1, 24°, 68° and 90°, and class 8, 27°, 39°, 46°, 55°, 65° and 66°) in the various classes is permissible only where classification under an item containing a specific heading is not possible.

Examples for the classification of mixtures and solutions in classes and items:

1. A mixture consists of a Class 3 (c) inflammable liquid, a Class 6.1 (b) toxic substance and a Class 8 (a) corrosive substance. Reading along the row from 3 (c) in the first column, the point of coincidence with the column headed 6.1 (b) shows that the cumulative risk for the first two substances is 6.1 (b). Then go to the row headed 6.1 (b); the point of coincidence with the column headed 8 (a) shows that the cumulative risk for the three substances is 8 (a). The mixture should, therefore, be assigned to Class 8 (a).
2. A solution of phenol of class 6.1, 13° (b) and benzene of class 3, 3° (b), is placed in class 3, letter (b); because of the toxicity of phenol, the solution is to be placed in class 3 under 17° (b).
3. A mixture of sodium arsenate of class 8, 41° (b), should be placed in class 6.1 under 51° (b).
4. A solution of naphtalene of class 4.1, 11° (b), in petrol of class 3, 3° (b), should be placed in class 3, 3° letter (b).

In summarizing the following points are remarkable:

1. Referring to all classes the following note was added:
   Note: For the classification of solutions and mixtures (such as preparations and wastes), see also marginal 3 (3) RID/2002 (8) ADR.

2. The inclusion of special requirements such as 314 (1) RID/2314 (1) ADR:
   For the carriage of wastes (see marg. 3 (4) RID/2000 (4) ADR) the description of the goods shall be:
   "Waste, containing ...", the component(s) which has/have been used for the classification of wastes under marginal 3 (3) RID/2002 (8) ADR to be entered under its/their chemical name(s), e.g. "Wastes, containing methanol, class 3, 17° (b), RID/ADR".

In general, not more than the two components which most predominantly contribute to the danger(s) of a mixture need to be shown.

3. Items to classify wastes in the manner of the N.O.S.-positions
   (N.O.S. = Not Otherwise Specified) of the UN-Recommendations
   "Transport of Dangerous Goods" are created; e.g.

   item 46   for "basic inorganic compounds or basic solu-
             tions or mixtures of inorganic substances" or
   item 55   for "basic organic compounds or basic solutions
             and mixtures of organic materials.

These wastes should be classified into the danger letters (a), (b)
and (c).

4. With regard to these "N.O.S."-positions and in case it is difficult to identify the waste, Schedule II states that the different ingredients, compositions and properties of the
   dangerous wastes, should be taken into account.

5. The amendment to the RID/ADR-Regulations also include
   packaging, labelling etc.

6. For soil contamined with e.g. solvents (a class 3 substance)
   which may be derived from a soil sanitation project, etc., a
   special number 1b has been introduced in the class 4.1
   (inflammable solids).
   Situations are imaginable, that soil is contamined with oil,
   which has a flashpoint > 100°C. This type of oil does not need
   a RID/ADR classification. In mixture with a solid (e.g. soil)
   it is quite possible that the flashpoint is lower than 100°C.
   Therefore it has to be classified into class 4.1 under number
   1b. It should be noticed, that the classification criteria in
   this case are not based on concentration but on inflammability.

7. It is expected that a class 9 will be introduced for substances
   which contains a danger to the environment (e.g. PCB and
   asbestos).

ANNEX IV

Listing of the toxic or dangerous wastes, according the Council
Directive 78/319/EEC in the classes as set out in the UN Recommen-
dations, RID, ADR and the IMO-code.
(Ref. annex IX/1f)

Toxic or dangerous waste (+)	Listing according to:			Notes
	UN	RID/ADR	IMO	
1. Arsenic; arsenic compounds				
Arsenic	6.1		6.1	
Liquid and solid compounds	6.1	6.1	6.1	
2. Mercury; mercury compounds				
Mercury	8		8	
Liquid and solid compounds	6.1	6.1	6.1	
3. Cadmium; cadmium compounds				
Cadmium				
Liquid and solid compounds	6.1	6.1	6.1	
4. Thallium; thallium compounds				
Thallium				
Liquid and solid compounds	6.1	6.1	6.1	
5. Beryllium; beryllium compounds				
Beryllium	6.1	6.1	6.1	
Liquid and solid compounds	6.1	6.1	6.1	
6. Chromium VI compounds				
Chromic anhydride	5.1	5.1	5.1	
Chromic acid solution	8	8	8	
Ammonium bichromate	5.1		5.1	
7. Lead; lead compounds				
Lead dioxide	5.1	5.1	5.1	
Lead alkyls	6.1	6.1	6.1	
Sulphates and muds with over 3% available sulphuric acid	8	8	8	
Muds with less than 3% available sulphuric acid		6.1		
Other compounds	6.1	6.1	6.1	7.1

Toxic or dangerous waste (+)	Listing according to:			Notes
	UN	RID/ADR	IMO	
8. Antimony; antimony compounds				
Antimony	6.1		6.1	
Tri- and pentachloride, penta-	8	8	8	
fluoride	6.1	6.1	6.1	
Other compounds				
9. Phenols; phenol compounds				
Phenol, chlorophenols, nitro-				
phenols	6.1	6.1	6.1	
Chlorophenates	8		8	
10. Cyanides, organic and inorganic				
Hydrocyanic acid absorbed in a				
porous inert material contai-				
ning not more than 20% acid	6.1	6.1	6.1	
Inorganic cyanides	6.1	6.1	6.1	
Organic cyanides	3; 6.1	3; 6.1	3; 6.1	10.1
11. Isocyanates	3; 6.1	3; 6.1	3; 6.1	11.1
12. Organic-halogen compounds, ex-				
cluding inert polymers and				
other substances referred to				
in this list	3; 6.1	3; 6.1	3; 6.1	12.1
13. Chlorinated solvents	3; 6.1	3; 6.1	3; 6.1	13.1
14. Organic solvents	3; 6.1	3; 6.1	3; 6.1	14.1
15. Biocides & phyto-pharmaceutical				
substances				
Sodium and potassium chlorate	5.1	5.1	5.1	
Inorganic weedkillers, chlo-				
rate-based	5.1	5.1	5.1	
Liquid pesticides	3; 6.1	3; 6.1	3; 6.1	15.1
Solid pesticides	6.1	6.1	6.1	
16. Tarry materials from refining				
and tar residues from distilling	3	3	3	

Toxic or dangerous waste (+)	Listing according to:			Notes
	UN	RID/ADR	IMO	
17. Pharmaceutical compounds				17.1
18. Peroxides, chlorates, perchlorates, azides				
Hydrogen peroxide	5.1	8; 5.1	5.1	18.1
Other inorganic peroxides	5.1	5.1	5.1	
Organic peroxides	5.2	5.2	5.2	18.2
Chlorates, perchlorates	5.1	5.1	5.1	
Barium azide wetted with not less than 50% water, by weight	4.1	6.1	4.1	18.3
Sodium azide	6.1	6.1	6.1	
Other azides				18.4
19. Ethers	3	3	3	
20. Non-identifiable and/or new laboratory chemicals whose effects on the environment are not known				20.1
21. Asbestos (dust and fibres)	9		9	
22. Selenium; selenium compounds				
Selenium, seleniates, selenites	6.1	6.1	6.1	
Selenic acid	8	8	8	
Selenium oxychloride	8		8	
23. Tellurium; tellurium compounds				
Tellurides		6.1		
24. Aromatic polycyclic compounds (with carcinogenic effects)				24.1
25. Metal carbonyls				
Ferro-pentacarbonyl, nickel-tetracarbonyl	6.1	6.1	6.1	

Toxic or dangerous waste (+)	Listing according to:			Notes
	UN	RID/ADR	IMO	
26. Soluble copper compounds				
Copper chloride	8	6.1		
Cupriethylenediamine solution	8	8	8	
27. Acids and/or basic substances used in the surface treatment and finishing of metals	8	8	8	
28. Polychlorinated biphenyls and triphenyls and mixtures thereof	9		9	

(+) The toxic and dangerous wastes referred to in Italian legislation, are listed in this column, numbered 1 to 28. Under each heading are, as appropriate, listed the dangerous substances as they appear in various international transport regulations, where it has been possible to insert them.

Notes:

7.1 In the case of UN and IMO, only if soluble.

10.1)

11.1) These are Class 3 if the flash   point is lower than 23°C

12.1) in the case of UN and IMO, and lower than 21°C in the

13.1) case of RID/ADR.

14.1)

15.1)

17.1) Classified according to the type of risk for each

20.1) individual substance.

24.1)

18.1 According to RID/ADR, solutions are Class 5.1 if they contain more than 60% hydrogen peroxide.

18.2 According to RID, organic peroxides may not be transported if their temperature has to be monitored during transport.

18.3 If barium azide is wetted with less than 50% water, it may not be transported according to RID/ADR; in the case of UN and IMO it is relisted as Class 1.

18.4 These may not be transported unless they are listed by name in Class 1).

ANNEX V

Physical, chemical and biological characteristics used as criteria
for assigning substances to hazard groups

The harmonized RID/ADR-regulations are conforming with the Recommendations of the UN-Committee of Experts.
Based on characteristics that are of relevance to hazards, the
recommendations state that there are three Packing Groups. These
Groups are:

Group I   (UN) = Group (a) RID/ADR; representing great danger
Group II (UN) = Group (b) RID/ADR; representing medium danger
Group III (UN) = Group (c) RID/ADR; representing minor danger

Dangerous goods within the various classes or divisions have been
assigned to three Packing Groups, except for the substances (with
which we are not concerned in this report) in:

Class 1 , all substances which have been assigned to Packing
Group II, although a packing system has been stated for each indi-
          vidual substance;

Class 2 , because of the special characteristics required of packa-
          gings for substances (gases) in this Class;

Class 7 , because of the hazards presented by substances in this
          Class (radioactive).

The criteria for the assignment of various goods to Packing
Classes have been laid down only for goods in Class 3, Class 6,
Division 6.1 and Class 8.

CRITERIA FOR CLASS 3
Goods in this Class having a flash point (in the closed-cup test)
equal to or less than 60.5°C, are assigned to:

Packing Group I if their initial boiling point is equal to or less
than 35°C;

Packing Group II if their initial boiling point is over 35°C and
their flash point (closed-cup) is less than 23°C;

Packing Group III if their initial boiling point is over 35°C and
their flash point (closed-cup) is equal to or greater than 23$_{\circ}$C.

Viscous substances (such as paints, varnishes and adhesives) with a flash point of less than 23°C are included in Packing Group III provided that:

a) less than 3% in volume of the clear solvent layer separates in the solvent separation test;

b) the mixture contains not more than 5% of substances in Group I or II of Division 6.1 or Class 8, or not more than 5% of substances in Group I of Class 3 requiring a Division 6.1 or Class 8 subsidiary label;

c) the capacity of the respectacle used is not more than 30 litres

d) the viscosity and flash point are in accordance with the following table:

Flowtime in seconds		Flash point in Degrees C
cup with 4 mm jet	cup with 8 mm jet	
> 20	-	17
> 60	-	10
> 100	-	5
> 160	-	-1
> 220	> 17	-5
	> 40	no lower limit

CRITERIA FOR CLASS 6, DIVISION 6.1

The criteria for assignment of goods in this Division to Packing Groups are based on oral toxicity, dermal toxicity and inhalation toxicity.

With reference to chapter 6 of the UN-Recommendations "Transport of Dangerous Goods" the definitions and criteria in the RID/ADR-Regulations, marginal 600/(2600) are in conformity.

With exception of substances in items 1-3 the substances are grouped into:

UN-Group I   = ADN/RID-group (a) very severe risk of
               poisoning
UN-Group II = ADN/RID-group (b) serious risk of poisoning
UN-Group III = ADN/RID-group (c) harmful, with relatively
               low risk of poisoning.

The substances are listed in the RID/ADR-Regulations in marginal 601/(2601). If classified substances of Class 6.1 contain components or impurities with other poisoning properties or another boiling point, their mixtures or solutions shall be classified with these items or letters, which are in accordance with their real toxicity or boiling point. If due to these components the flashpoint comes below 21°C, the substance shall be classified according to their toxicity into class 3 and respective items and letters.

If substances of class 6.1 have predominantly corrosive properties due to components or impurities of substances of class 8, their mixtures and solutions shall be classified according to their items and letters into class 8.

When using this grouping (a-c), account should be taken of past experience in instances of accidental poisoning and of special properties possessed by any individual substance, such as liquid state, high volatility, any special likelihood of penetration, and special biological effects.

In the absence of human experience the grouping should be based on data from animal experiments, as shown in the tables below:

Table: Grouping criteria for oral and dermal contact

Group in letter	Oral toxicity/LD_{50}(mg/kg)	Dermal toxicity/LD_{50}(mg/kg)
(a)	<5	<40
(b)	>5 - 50	>40 - 200
(c)	solids >50 - 500 liquids >50 - 2000	>200 - 1000

Three methods were developed for the criteria for inhalation of toxic vapours, dusts or mists, a combination of LC_{50} and volatility:

Table: Grouping criteria for inhalation of toxic vapours, dusts and mists

Group in letter	vapours			dusts and mists
	Toxid point method	Boiling point method	Vapour concentration method	LC_{50} (mg/l)
	A	B	C	
(a)	2,7	4,5	50	0,5
(b)	2,7 - 3,9	4,5 - 5,7	50 - 500	0,5 - 2
(c)	3,9 - 5,1	5,7 - 6,9	500 - 5000	2 - 10

CRITERIA FOR CLASS 8

The criteria for the allocation of Class 8 goods to Packing Groups
are based on the length of contact necessary to produce visible
necrosis in human skin:

Depending on the danger, the test criteria for the three groups in this
class in the UN-Recommendations are:

Group I (very dangerous substances):
substances that cause visible necrosis of the skin tissue at the site
of contact when tested on the intact skin of an animal for a period of
not more than 3 minutes.

Group II (substances presenting medium danger):
substances that cause visible necrosis of the skin tissue at the site
of contact when tested on the intact skin of an animal for a period of
more than 3 but not more than 60 minutes.

Group III (substances presenting minor danger):
substances that cause visible necrosis of the skin tissue at the site
of contact when tested on the intact skin of an animal for a period of
not more than 4 hours.

Regarding Class 8, marginal 800/2800, footnote 1 of RID/ADR conforms
with these recommendations; with exception of the group terms (the differently designated group terms are valid for all classes of
RID/ADR).

GOODS PRESENTING MORE THAN ONE RISK

The following table is used to determine the Group of a class of
substance having more than one risk, when it is not named in the
UN-list of dangerous goods most commonly carried.

	4.2*	4.3*	5.1 I*,**	5.1 II*,**	5.1 III*,**	6.1 I (Inh)	6.1 I (Derm)	6.1 I (Oral)	6.1 II	6.1 III	8 IL	8 Is	8 IIL	8 IIg	8 IIIL	8 IIIg
3 I.......			3	3	3	6.1	3	3	3	3	3	-	3	-	3	-
3 II......			3	3	3	6.1	3	3	3	3	3	-	3	-	3	-
3 III.....			3	3	3	6.1	3	3	3	3	3	-	3	-	3	-
4.1 I*....	4.2	4.3	4.1	4.1	4.1	6.1	6.1	4.1	6.1	4.1	4.1	4.1	4.1	4.1	4.1	4.1
4.1 II*...	4.2	4.3	4.1	4.1	4.1	6.1	6.1	4.1	6.1	4.1	8	4.1	8	4.1	4.1	4.1
4.1 III*..	4.2	4.2	4.1	4.1	4.1	6.1	6.1	4.1	6.1	4.1	8	-	8	-	4.2	4.2
4.2 I*....		4.2	4.2	4.2	4.2	6.1	6.1	4.2	6.1	4.2	4.2	4.2	4.2	4.2	4.2	4.2
4.2 II*...		4.2	5.1	4.2	4.2	6.1	6.1	4.2	6.1	4.2	8	4.2	8	4.3	4.2	4.2
4.2 III*..		4.3	5.1	5.1	4.2	6.1	6.1	4.2	6.1	4.3	8	4.3	8	4.3	8	4.3
4.3 I*....			5.1	5.1	4.3	6.1	6.1	4.3	6.1	4.3	8	4.3	8	4.3	8	4.3
4.3 II*...			5.1	5.1		6.1	6.1	4.3	6.1	4.3	8	4.3	8	8	8	8
4.3 III*..			5.1	5.1		6.1	6.1	4.3	6.1	4.3	8	8	8	8	8	8
5.1 I*....						6.1	6.1	5.1	6.1	5.1	5.1	5.1	5.1	5.1	5.1	5.1
5.1 II*...						6.1	6.1	5.1	6.1	5.1	8	6.1	8	6.1	5.1	5.1
5.1 III*..						6.1	6.1	6.1	6.1	6.1	8	6.1	8	6.1	6.1	6.1
6.1 I (Inh)..											6.1	6.1	6.1	6.1	6.1	6.1
6.1 I (Derm).											6.1	6.1	6.1	6.1	6.1	6.1
6.1 I (Oral).											6.1	6.1	6.1	6.1	6.1	6.1
6.1 II (Inh).											6.1	6.1	6.1	6.1	6.1	6.1
6.1 II (Derm)											6.1	6.1	6.1	6.1	6.1	6.1
6.1 II (Oral)											6.1	6.1	6.1	6.1	6.1	6.1
6.1 III....											8	8	8	8	8	8

* There are at present no established criteria for determining Packing Groups within Class 4, Divisions 1, 2 and 3, and Class 5, Division 1. For the time being, the degree of hazard is to be assessed by analogy with listed substances, allocating the substances to I, high; II, medium; and III, low risk.

** The precedence of hazard characteristics of substances with an oxidizing component is given only as a guide. Since the combination will give rise to increased reactivity, each substance should be considered individually.

- Denotes an impossible combination.

The precedence of hazard characteristics of substances in Classes 1, 2, 7 and Division 5.2 has not been dealt with, since these primary characteristics take precedence.

ANNEX VI

Significant limits for drivers training and licensing

The driver training required in the ADR-regulations should be the basis for gaining knowledge of the properties of hazardous materials and dangers in their transportation. Dangerous wastes can contain small quantities of substances which, under normal transport conditions, would present only a minor degree of danger, but which in case of an accident may be the unexpected source of the development of serious dangers for the vicinity.

The relevant driver training requirements of marg. 10 315/ADR*) must be seen in connection with marg. 10 011/ADR and the various included tables. These tables give a survey on the limited quantities of dangerous goods which can be transported in one unit without special driver rtaining and licensing requirements (only in transfrontier shipment).

In the case of the conveyance of a mixture of dangerous wastes, the following examples explain the conroversing limits of these requirements and the dangers which can develop when transporting quantities below these limits.

The following table gives a survey of the properties and limited quantities of some substances of Class 3 (flammable liquids):

properties and limited quantities

substance	boiling point °C	vapour pressure mbar/°C	rel.den-sity to air/20°C	flash point °C	ignition temperat. °C	explo.limit volume-% in air	limited quantities in kg
Diethylether	34,6	853/30	2,56	below-20	180	1,7 - 36	333
Isoprene	34,3	933/30	2,35	" -20	220	1 - 9,7	333
Iso-Pentane	28	1100/30	2,49	" -20	230	1,8 - 7,6	333
Ethylacetate	72	148/30	3,04	-4	460	2,1 - 11,5	333
m-Xylol	139	147/30	3,67	25	525	1,1 - 7	500

Nitrocellulose (item 34 c), a mixture with liquids of item 32 c (flash point = 55 - 100°C) with not exceeding 55% Nitrocellulose with not more than 12,6% Nitrogen (e.g. nitro lacquers)

---

*) A driver of tank cars or transport units for the conveyance of tanks or tank-containers with a volume above 3 000 litres must prove the successful comletion of such a driver training course concerning dangerous substances

Some examples of mixtures can explain these problems:

325 kg Ethylacetate (EA) at 20°C may develop a vapour cloud of 88 m^3. The lower explosible limit is 2,1 % in air = 4 200 m^3 explosible atmosphere. The cloud is heavier than air and can "creep" from the outlets to distant areas (basements) and, in relation to the density, may reach a height of approximately 5 meters. The result is a dangerous area of 29 x 29 m. If this incident occurs on a street (10 m wide/houses higher than 8 m) the cloud has a length of 84 m. Nitrocellulose (NC) gives off toxic fumes (nitric oxides) in case of a fire, which cause pulmonary edema (100 ppm; lethal after 5 hours). Thus, the original uncomplicated mixture could be the source of an unexpected catastrophic incident, as shown on the following table:

unit of transportation	vapour in m^3	atmosphere in m^3 explosible	toxic	height of dang.cloud	area of danger in m^2	dangerous situation in street(length in m)
324 kg (EA)	88	4 220	-	5 m	29 x 29	84
25 kg (NC)	6	-	60 000	5 m	109 x 109	1 200
450 kg Xylol	25	2 270	-	4 m	24 x 24	64
50 kg (NC)	12	-	120 000	5 m	155 x 155	2 400

These considerations should be the basis for a proposal regarding a revision of the driver training and licensing requirements concerning the transport of dangerous wastes.

ANNEX VII

Emergency action codes: Kemler system and Hazchem system

Kemler code:
For dangerous products being transported by road or rail, the use
of the Kemler number is internationally accepted. It consists of
two or three digits of which the first indicates the main hazard
of the substance, such as:
2. Gas (pressurized, in solution or refrigerated)
3. Flammable liquid
4. Flammable solid
5. Oxidizing substance or organic peroxide
6. Toxic
8. Corrosive

The second and the third digit provide subsidiary information
regarding the hazard of the material, such as:
0. No subsidiary hazard
1. Potentially explosive
2. Risk for gaseous emissions
3. Flammable
5. Oxidizing
6. Toxic
8. Corrosive
9. Risk for violent reaction, e.g. spontaneous decomposition or
   polymerization

The combinations of digits may have specific meanings, such as:
333   auto-ignitable liquid
 42   solid generating gas on contact with water
 44   flammable solid in the melted state
 22   refrigerated gas
 33   very flammable liquid
 66   very toxic
 55   very corrosive

The identification number being preceded by "x" means that the substance will react violently on contact with water. It may be an indication for the use of particular fire extinguishing materials.

Hazchem code:
The Hazchem code serves as a source of immediate information for the fire service and police (Hazchem is developed by the London fire service); this information is about the appropriate emergency actions to be taken by the fire service or police. An example of an Hazchem code: 2PE.
The following Hazchem scale is given by Heinz Dorias in "Gefährliche Güter" (1984, Springer-Verlag, Berlin).

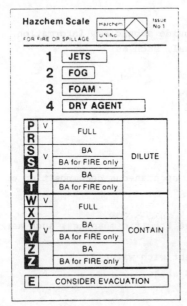

Hazchem Scale	Hazchem UN No	Issue No 1
FOR FIRE OR SPILLAGE		

1 JETS
2 FOG
3 FOAM
4 DRY AGENT

P	V	FULL	
R			
S	V	BA	DILUTE
S		BA for FIRE only	
T		BA	
T		BA for FIRE only	
W	V	FULL	
X			
Y	V	BA	CONTAIN
Y		BA for FIRE only	
Z		BA	
Z		BA for FIRE only	

E	CONSIDER EVACUATION

Notes for Guidance

FOG
In the absence of fog equipment a fine spray may be used.

DRY AGENT
Water **must not** be allowed to come into contact with the substance at risk.

V
Can be violently or even explosively reactive.

FULL
Full body protective clothing with BA.

BA
Breathing apparatus plus protective gloves

DILUTE
May be washed to drain with large quantities of water.

CONTAIN
Prevent, by any means available, spillage from entering drains or water course.

Printed in England for Her Majesty's Stationery Office by Robendene (Chesham) Ltd.
Dd. 506846 K1200 5/75
3p net.
ISBN 0 11 340752 1

ANNEX VIII

The national reports on the Transport of Non-Nuclear Toxic and Dangerous Wastes (technical, safety and legal aspects) were prepared by

- ir. Jan M. Willocx, Londerzeel (Belgian report)

- Dr. Heinz Dorias, Bremerhaven (German report)

- Roger W. Short et al.,Jydsk Teknologisk Institut, Århus
  (Danish report)

- P. Vincent, SCORI, Bois d'Arcy (French report)

- Matthew Linch, IIRS, Dublin (Irish report)

- Dr. Lidia Failla et al., INPRAT, Rome (Italian report)

- ir. Harry M. Kaspers et al., TAUW Infra Consult B.V. Deventer
  (Dutch report)

- ir. Jan M. Willocx, Londerzeel (Luxembourg report)

- Dr. T. Norton, Newcastle-upon-Tyne (U.K. report)

EUROPEAN RESEARCH & CONSULTING sprl

TRANSPORT OF NON-NUCLEAR TOXIC AND DANGEROUS WASTES

technical, safety and legal aspects especially regarding packaging
and means of transport

FINAL CHECK-LIST

I. Classification of dangerous wastes - (substances), technical
and identification aspects

1. Definition of dangerous wastes according to the respective
national system

2. Identification of the particular classes of risk, considering

2.1.1 "Solutions", means any homogeneous liquid mixture of two or
more chemical compounds or elements that will
not undergo any segregation under conditions normal to transportation

2.1.2 "Mixtures", means heterogeneous materials composed of more than
one chemical compound or element in the same or
different aggregation

2.2 Criteria approach

2.2.1 Physical and chemical properties:

2.2.1.1 melting point

2.2.1.2 boiling point

2.2.1.3 vapour pressure

2.2.1.4 flashpoint

2.2.1.5 flammability

2.2.1.6 explosive properties

2.2.1.7 auto-flammability (spontaneous ignition)

2.2.1.8 chemical reaction with water

2.2.1.9 oxidising properties

2.2.1.1o unstable contents

2.2.1.11 density

2.2.1.12 viscosity

2.2.1.13 corrosivity (to materials)

2.2.2   Health hazards

2.2.2.1 irritation (skin, eyes, mucuous membranes etc)

2.2.2.2 acute toxicity (orally, by inhalation, skin absorption)

2.2.2.3 chronic toxicity

2.2.2.4 genetic effects (carcinogenity, mutagenicity, teratogenicity)

2.2.3   Other criteria approaches

2.3   Hazard analysis

2.3.1   main components of solutions and mixtures with respect to hazards

2.3.2   properties of materials in 2.3.1

2.3.3   classifying into classes, special digits and (packing-)groups of danger

2.3.4   other methods of disposal distribution

2.4   List approach

2.4.1   list of substances or chemicals which are considered as dangerous waste

2.4.2   list of activities or industrial processes of which waste is considered as dangerous waste

2.4.3   List of excepted substances and/or activities of industrial processes (e.g. sewage sludge in NL)

2.5   Particular criteria for manipulation

2.5.1   pumpability (suitability for pumping)

2.5.2   dumping ability (suitability for landfill)

2.5.3   suction ability (e.g. suitable for suction pumps)

2.5.4   other possibilities of manipulation

2.6   Special classifying systems (methods : e.g. IMO, Pavony, FRG)

3.   Competence (authorities, administrations, test facilities etc.) and procedures

3.1   legislation and Codes of Practice (for wastes)

3.2   new methods of determination of classes of risk

3.3   transportation

3.4   test facilities

4.   State of the art and research in the member states with regard to

4.1   existing procedures

4.2   new methods of determination of classes of risk

5.   Relationship between international conventions (e.g.ADR) and
       national law (esp.supplements,derogations)

II.   Packaging

1.   Types and conditions
1.1   general packaging requirements
1.2   types of packaging (codes and specifications):
1.2.1 packaging with a capacity not exceeding 450 l or 400 kg net
       weight
1.2.2 containers (capacity between 450 l and 3.000 l)
1.2.3 containers (capacity above 3.000 l)
1.2.4 tankcontainers
1.2.5 tanks

1.3   Consignment procedures

1.3.1 marking of packaging (e.g. class, division)
1.3.2 labels identifying risks
1.3.3 documentation of dangerous wastes shipment
1.3.4 placarding (e.g. orange placard)

1.4   performance tests of packagings (pos.1.2.1):

1.4.1 drop test
1.4.2 leakproofness test
1.4.3 internal pressure (hydraulic) test
1.4.4 stacking test
1.4.5 corrosion resistance

1.5   container performance tests

1.5.1 lifting test
1.5.2 stacking test
1.5.3 top (roof) load test
1.5.4 bottom load test
1.5.5 front and rear wall test
1.5.6 (distortion) rigidity test
1.5.7 side wall tests

1.6   performance tests for tankcontainer (IMO-types I,II,IV,V;
       RID/App.X; ADR/App.B1b)

1.7   regulations for tanks (RID/App.XI; ADR/App. B1a)

1.8   <u>special tankers and attached tanks</u>

1.8.1   special containersfor transportation of wastes

1.8.2   special pressure-vacuum-tanks, e.g. for road and rail (tankcars)

1.8.3   regulations for containers in pos. 1.8.1 and 1.8.2

1.8.3.1 general requirements

1.8.3.2 construction

1.8.3.3 equipment

1.8.3.4 performance testing

1.8.3.5 marking

1.8.3.6 operating procedures

1.8.3.7 special regulations

1.9   <u>cleaning of packagings</u>

1.9.1   methods and practices for cleaning small receptacles (included e.g. 1.2.1)

1.9.2   methods and practices for cleaning containers and tanks

2.   <u>State of the art and research with regard to the safety of containers for the transport of dangerous wastes (substances)</u>

3.   <u>Applicability of special test procedures by container manufacturers for transport container</u>

4.   <u>Relationship between international conventionsand national law</u>

III.   <u>Means of transport</u> (road and rail vehicles, sea-going and inland waterway vessels)

1.   <u>Road vehicles</u>

1.1   <u>general requirements</u>

1.2   <u>types of road vehicles</u>

1.3   <u>special requirements</u>

1.3.1   conveyance in receptables, whether on pallets or not

1.3.2   conveyance in containers other than tanks or tankcontainers

1.3.3   conveyance in tankcontainers

1.3.4   conveyance in tanks

1.4   <u>electrical devices and systems</u>

1.5   <u>fire extinguishing equipment</u>

303

2.1   covered loads

2.1.1 receptacles held or carried on pallets

2.1.2 receptacles without pallets

2.2   in bulk

2.3   in containers

2.4   in tankcontainers

2.5   in tanks

2.6   in combined packaging

3.   Loading and unloading equipment

4.   Safety measures

5.   Special safety measures for special loads (e.g. receptacles for liquid wastes)

6.   Relationship between international conventions and national law

V.   Control and monitoring systems

1.   General requirements

2.   Listing and description of existing measuring and testing procedures for:

2.1   different dangerous wastes as outlined in I.2.1.1 and 2.1.2

2.2   safety in traffic (vehicles and vessels)

2.3   ancillary equipment

2.4   packaging, especially :

2.4.1 materials

2.4.2 corrosion

2.4.3 leakage

2.5   loading and unloading equipment

2.6   loads

2.6.1 sampling

2.6.2 analysis based on spot checks of key substances

3.   State of the art and research

4.   Procedures in the event of failure

5. International conventions

VI.   <u>Use of electronic data processing systems for the organisation and supervision of the transfrontier shipment of dangerous wastes (substances)</u>

1.   By road

2.   By rail

3.   By inland waterways

4.   In coastal waters or on the high seas

VII.   <u>Emergency measures</u>

1.   Planned measures in the event of accidents during the transport-ation of dangerous wastes (substances) to minimise environmental damage

2.   Emergency and consultation centres

3.   Technical emergency organisations

4.   National alarm and disaster plans

5.   Transport Emergency Cards

6.   Emergency action code

7.   Driver training and licensing

8.   International conventions

VIII. <u>Evaluation of respective national systems, including possible shortcomings and suggested alternative solutions</u>

Transport of Dangerous Wastes

Luxembourg: Office for Official Publications of the European Communities

87 - 316 pp.   240 x 173 cm

EN

ISBN 92-825-6748-6

Catalogue number: SY-48-87-234-EN-C

Price (excluding VAT) in Luxembourg:

ECU 23.20   BFR 1.000   ULK 16.80   USD 25.00   IRL 17.80

Venta y suscripciones · Salg og abonnement · Verkauf und Abonnement · Πωλήσεις και συνδρομές
Sales and subscriptions · Vente et abonnements · Vendita e abbonamenti
Verkoop en abonnementen · Venda e assinaturas

BELGIQUE / BELGIË

Moniteur belge / Belgisch Staatsblad
Rue de Louvain 40-42 / Leuvensestraat 40-42
1000 Bruxelles / 1000 Brussel
Tél. 512 00 26
CCP / Postrekening 000-2005502-27

Sous-dépôts / Agentschappen:

**Librairie européenne /
Europese Boekhandel**
Rue de la Loi 244 / Wetstraat 244
1040 Bruxelles / 1040 Brussel

CREDOC
Rue de la Montagne 34 / Bergstraat 34
Bte 11 / Bus 11
1000 Bruxelles / 1000 Brussel

DANMARK

Schultz EF-publikationer
Møntergade 19
1116 København K
Tlf: (01) 14 11 95
Telecopier: (01) 32 75 11

BR DEUTSCHLAND

Bundesanzeiger Verlag
Breite Straße
Postfach 10 80 06
5000 Köln 1
Tel. (02 21) 20 29-0
Fernschreiber: ANZEIGER BONN 8 882 595
Telecopierer: 20 29 278

GREECE

G.C. Eleftheroudakis SA
International Bookstore
4 Nikis Street
105 63 Athens
Tel. 322 22 55
Telex 219410 ELEF

Sub-agent for Northern Greece:

Molho's Bookstore
The Business Bookshop
10 Tsimiski Street
Thessaloniki
Tel. 275 271
Telex 412885 LIMO

ESPAÑA

Boletín Oficial del Estado
Trafalgar 27
28010 Madrid
Tel. (91) 446 60 00

Mundi-Prensa Libros, S.A.
Castelló 37
28001 Madrid
Tel. (91) 431 33 99 (Libros)
        431 32 22 (Suscripciones)
        435 36 37 (Dirección)
Télex 49370-MPLI-E

FRANCE

Journal officiel
**Service des publications
des Communautés européennes**
26, rue Desaix
75727 Paris Cedex 15
Tél. (1) 45 78 61 39

IRELAND

Government Publications Sales Office
Sun Alliance House
Molesworth Street
Dublin 2
Tel. 71 03 09

or by post

Government Stationery Office
Publications Section
6th floor
Bishop Street
Dublin 8
Tel. 78 16 66

ITALIA

Licosa Spa
Via Lamarmora, 45
Casella postale 552
50 121 Firenze
Tel. 57 97 51
Telex 570466 LICOSA I
CCP 343 509

Subagenti:

Libreria scientifica Lucio de Biasio - AEIOU
Via Meravigli, 16
20 123 Milano
Tel. 80 76 79

Libreria Tassi
Via A. Farnese, 28
00 192 Roma
Tel. 31 05 90

Libreria giuridica
Via 12 Ottobre, 172/R
16 121 Genova
Tel. 59 56 93

GRAND-DUCHÉ DE LUXEMBOURG
et autres pays / and other countries

**Office des publications officielles
des Communautés européennes**
2, rue Mercier
L-2985 Luxembourg
Tél. 49 92 81
Télex PUBOF LU 1324 b
CCP 19190-81
CC bancaire BIL 8-109/6003/200

Abonnements / Subscriptions

Messageries Paul Kraus
11, rue Christophe Plantin
L-2339 Luxembourg
Tél. 49 98 888
Télex 2515
CCP 49242-63

NEDERLAND

Staatsdrukkerij- en uitgeverijbedrijf
Christoffel Plantijnstraat
Postbus 20014
2500 EA 's-Gravenhage
Tel. (070) 78 98 80 (bestellingen)

PORTUGAL

**Imprensa Nacional
Casa da Moeda, E. P.**
Rua D. Francisco Manuel de Melo, 5
1092 Lisboa Codex
Tel. 69 34 14
Telex 15328 INCM

Distribuidora Livros Bertrand Lda.
Grupo Bertrand, SARL
Rua das Terras dos Vales, 4-A
Apart. 37
2700 Amadora CODEX
Tel. 493 90 50 - 494 87 88
Telex 15798 BERDIS

UNITED KINGDOM

HM Stationery Office
HMSO Publications Centre
51 Nine Elms Lane
London SW8 5DR
Tel. (01) 211 56 56

Sub-agent:

Alan Armstrong & Associates Ltd
72 Park Road
London NW1 4SH
Tel. (01) 723 39 02
Telex 297635 AAALTD G

UNITED STATES OF AMERICA

**European Community Information
Service**
2100 M Street, NW
Suite 707
Washington, DC 20037
Tel. (202) 862 9500

CANADA

Renouf Publishing Co., Ltd
61 Sparks Street
Ottawa
Ontario K1P 5R1
Tel. Toll Free 1 (800) 267 4164
Ottawa Region (613) 238 8985-6
Telex 053-4936

JAPAN

Kinokuniya Company Ltd
17-7 Shinjuku 3-Chome
Shiniuku-ku
Tokyo 160-91
Tel. (03) 354 0131

Journal Department
PO Box 55 Chitose
Tokyo 156
Tel. (03) 439 0124